Big Data Analytics with Spark

A Practitioner's Guide to Using Spark for Large-Scale Data Processing, Machine Learning, and Graph Analytics, and High-Velocity Data Stream Processing

Mohammed Guller

Apress®

Big Data Analytics with Spark

ISBN-13 (pbk): 978-1-4842-0965-3

ISBN-13 (electronic): 978-1-4842-0964-6

Managing Director: Welmoed Spahr
Lead Editor: Celestin John Suresh
Development Editor: Chris Nelson
Technical Reviewers: Sundar Rajan Raman and Heping Liu
Editorial Board: Steve Anglin, Louise Corrigan, Jim DeWolf, Jonathan Gennick, Robert Hutchinson, Michelle Lowman, James Markham, Susan McDermott, Matthew Moodie, Jeffrey Pepper, Douglas Pundick, Ben Renow-Clarke, Gwenan Spearing
Coordinating Editor: Jill Balzano
Copy Editor: Kim Burton-Weisman
Compositor: SPi Global
Indexer: SPi Global
Artist: SPi Global

Distributed to the book trade worldwide by Springer Science+Business Media New York, 233 Spring Street, 6th Floor, New York, NY 10013. Phone 1-800-SPRINGER, fax (201) 348-4505, e-mail orders-ny@springer-sbm.com, or visit www.springer.com. Apress Media, LLC is a California LLC and the sole member (owner) is Springer Science + Business Media Finance Inc (SSBM Finance Inc). SSBM Finance Inc is a Delaware corporation.

For information on translations, please e-mail rights@apress.com, or visit www.apress.com.

Apress and friends of ED books may be purchased in bulk for academic, corporate, or promotional use. eBook versions and licenses are also available for most titles. For more information, reference our Special Bulk Sales–eBook Licensing web page at www.apress.com/bulk-sales.

Any source code or other supplementary material referenced by the author in this text is available to readers at www.apress.com. For detailed information about how to locate your book's source code, go to www.apress.com/source-code/.

To my mother, who not only brought me into this world, but also raised me with unconditional love.

Contents at a Glance

Contents

About the Author

Mohammed Guller is the principal architect at Glassbeam, where he leads the development of advanced and predictive analytics products. He is a big data and Spark expert. He is frequently invited to speak at big data–related conferences. He is passionate about building new products, big data analytics, and machine learning.

Over the last 20 years, Mohammed has successfully led the development of several innovative technology products from concept to release. Prior to joining Glassbeam, he was the founder of TrustRecs. com, which he started after working at IBM for five years. Before IBM, he worked in a number of hi-tech start-ups, leading new product development.

Mohammed has a master's of business administration from the University of California, Berkeley, and a master's of computer applications from RCC, Gujarat University, India.

About the Technical Reviewers

Sundar Rajan Raman is a big data architect currently working for Bank of America. He has a bachelor's of technology degree from the National Institute of Technology, Silchar, India. He is a seasoned Java and J2EE programmer with expertise in Hadoop, Spark, MongoDB, and big data analytics. He has worked at companies such as AT&T, Singtel, and Deutsche Bank. He is also a platform specialist with vast experience in SonicMQ, WebSphere MQ, and TIBCO with respective certifications. His current focus is on big data architecture. More information about Raman is available at https://in.linkedin.com/pub/sundar-rajan-raman/7/905/488.

I would like to thank my wife, Hema, and daughter, Shriya, for their patience during the review process.

Heping Liu has a PhD degree in engineering, focusing on the algorithm research of forecasting and intelligence optimization and their applications. Dr. Liu is an expert in big data analytics and machine learning. He worked for a few startup companies, where he played a leading role by building the forecasting, optimization, and machine learning models under the big data infrastructure and by designing and creating the big data infrastructure to support the model development.

Dr. Liu has been active in the academic area. He has published 20 academic papers, which have appeared in *Applied Soft Computing* and the *Journal of the Operational Research Society*. He has worked as a reviewer for 20 top academic journals, such as *IEEE Transactions on Evolutionary Computations* and *Applied Soft Computing*. Dr. Liu has been the editorial board member of *International Journal of Business Analytics*.

Acknowledgments

Many people have contributed to this book directly or indirectly. Without the support, encouragement, and help that I received from various people, it would have not been possible for me to write this book. I would like to take this opportunity to thank those people.

First and foremost, I would like to thank my beautiful wife, Tarannum, and my three amazing kids, Sarah, Soha, and Sohail. Writing a book is an arduous task. Working full-time and writing a book at the same time meant that I was not spending much time with my family. During work hours, I was busy with work. Evenings and weekends were completely consumed by the book. I thank my family for providing me all the support and encouragement. Occasionally, Soha and Sohail would come up with ingenious plans to get me to play with them, but for most part, they let me work on the book when I should have been playing with them.

Next, I would like to thank Matei Zaharia, Reynold Xin, Michael Armbrust, Tathagata Das, Patrick Wendell, Joseph Bradley, Xiangrui Meng, Joseph Gonzalez, Ankur Dave, and other Spark developers. They have not only created an amazing piece of technology, but also continue to rapidly enhance it. Without their invention, this book would not exist.

Spark was new and few people knew about it when I first proposed using it at Glassbeam to solve some of the problems we were struggling with at that time. I would like to thank our VP of Engineering, Ashok Agarwal, and CEO, Puneet Pandit, for giving me the permission to proceed. Without the hands-on experience that I gained from embedding Spark in our product and using it on a regular basis, it would have been difficult to write a book on it.

Next, I would like to thank my technical reviewers, Sundar Rajan Raman and Heping Liu. They painstakingly checked the content for accuracy, ran the examples to make sure that the code works, and provided helpful suggestions.

Finally, I would like to thank the people at Apress who worked on this book, including Chris Nelson, Jill Balzano, Kim Burton-Weisman, Celestin John Suresh, Nikhil Chinnari, Dhaneesh Kumar, and others. Jill Balzano coordinated all the book-related activities. As an editor, Chris Nelson's contribution to this book is invaluable. I appreciate his suggestions and edits. This book became much better because of his involvement. My copy editor, Kim Burton-Weisman, read every sentence in the book to make sure it is written correctly and fixed the problematic ones. It was a pleasure working with the Apress team.

—Mohammed Guller
Danville, CA

Introduction

This book is a concise and easy-to-understand tutorial for big data and Spark. It will help you learn how to use Spark for a variety of big data analytic tasks. It covers everything that you need to know to productively use Spark.

One of the benefits of purchasing this book is that it will help you learn Spark efficiently; it will save you a lot of time. The topics covered in this book can be found on the Internet. There are numerous blogs, presentations, and YouTube videos covering Spark. In fact, the amount of material on Spark can be overwhelming. You could spend months reading bits and pieces about Spark at different places on the Web. This book provides a better alternative with the content nicely organized and presented in an easy-to-understand format.

The content and the organization of the material in this book are based on the Spark workshops that I occasionally conduct at different big data–related conferences. The positive feedback given by the attendees for both the content and the flow motivated me to write this book.

One of the differences between a book and a workshop is that the latter is interactive. However, after conducting a number of Spark workshops, I know the kind of questions people generally have and I have addressed those in the book. Still, if you have questions as you read the book, I encourage you to contact me via LinkedIn or Twitter. Feel free to ask any question. There is no such thing as a stupid question.

Rather than cover every detail of Spark, the book covers important Spark-related topics that you need to know to effectively use Spark. My goal is to help you build a strong foundation. Once you have a strong foundation, it is easy to learn all the nuances of a new technology. In addition, I wanted to keep the book as simple as possible. If Spark looks simple after reading this book, I have succeeded in my goal.

No prior experience is assumed with any of the topics covered in this book. It introduces the key concepts, step by step. Each section builds on the previous section. Similarly, each chapter serves as a stepping-stone for the next chapter. You can skip some of the later chapters covering the different Spark libraries if you don't have an immediate need for that library. However, I encourage you to read all the chapters. Even though it may not seem relevant to your current project, it may give you new ideas.

You will learn a lot about Spark and related technologies from reading this book. However, to get the most out of this book, type the examples shown in the book. Experiment with the code samples. Things become clearer when you write and execute code. If you practice and experiment with the examples as you read the book, by the time you finish reading it, you will be a solid Spark developer.

One of the resources that I find useful when I am developing Spark applications is the official Spark API (application programming interface) documentation. It is available at `http://spark.apache.org/docs/latest/api/scala`. As a beginner, you may find it hard to understand, but once you have learned the basic concepts, you will find it very useful.

Another useful resource is the Spark mailing list. The Spark community is active and helpful. Not only do the Spark developers respond to questions, but experienced Spark users also volunteer their time helping new users. No matter what problem you run into, chances are that someone on the Spark mailing list has solved that problem.

And, you can reach out to me. I would love to hear from you. Feedback, suggestions, and questions are welcome.

—Mohammed Guller
LinkedIn: `www.linkedin.com/in/mohammedguller`
Twitter: @MohammedGuller

CHAPTER 1

■ ■ ■

Big Data Technology Landscape

We are in the age of big data. Data has not only become the lifeblood of any organization, but is also growing exponentially. Data generated today is several magnitudes larger than what was generated just a few years ago. The challenge is how to get business value out of this data. This is the problem that big data–related technologies aim to solve. Therefore, big data has become one of the hottest technology trends over the last few years. Some of the most active open source projects are related to big data, and the number of these projects is growing rapidly. The number of startups focused on big data has exploded in recent years. Large established companies are making significant investments in big data technologies.

Although the term "big data" is hot, its definition is vague. People define it in different ways. One definition relates to the volume of data; another definition relates to the richness of data. Some define big data as data that is "too big" by traditional standards; whereas others define big data as data that captures more nuances about the entity that it represents. An example of the former would be a dataset whose volume exceeds petabytes or several terabytes. If this data were stored in a traditional relational database (RDBMS) table, it would have billions of rows. An example of the latter definition is a dataset with extremely wide rows. If this data were stored in a relational database table, it would have thousands of columns. Another popular definition of big data is data characterized by three Vs: volume, velocity, and variety. I just discussed volume. *Velocity* means that data is generated at a fast rate. *Variety* refers to the fact that data can be unstructured, semi-structured, or multi-structured.

Standard relational databases could not easily handle big data. The core technology for these databases was designed several decades ago when few organizations had petabytes or even terabytes of data. Today it is not uncommon for some organizations to generate terabytes of data every day. Not only the volume of data, but also the rate at which it is being generated is exploding. Hence there was a need for new technologies that could not only process and analyze large volume of data, but also ingest large volume of data at a fast pace.

Other key driving factors for the big data technologies include scalability, high availability, and fault tolerance at a low cost. Technology for processing and analyzing large datasets has been extensively researched and available in the form of proprietary commercial products for a long time. For example, MPP (massively parallel processing) databases have been around for a while. MPP databases use a "shared-nothing" architecture, where data is stored and processed across a cluster of nodes. Each node comes with its own set of CPUs, memory, and disks. They communicate via a network interconnect. Data is partitioned across a cluster of nodes. There is no contention among the nodes, so they can all process data in parallel. Examples of such databases include Teradata, Netezza, Greenplum, ParAccel, and Vertica. Teradata was invented in the late 1970s, and by the 1990s, it was capable of processing terabytes of data. However, proprietary MPP products are expensive. Not everybody can afford them.

This chapter introduces some of the open source big data–related technologies. Although it may seem that the technologies covered in this chapter have been randomly picked, they are connected by a common theme. They are used with Spark, or Spark provides a better alternative to some of these technologies. As you start using Spark, you may run into these technologies. In addition, familiarity with these technologies will help you better understand Spark, which we will introduce in Chapter 3.

Hadoop

Hadoop was one of the first popular open source big data technologies. It is a scalable fault-tolerant system for processing large datasets across a cluster of commodity servers. It provides a simple programming framework for large-scale data processing using the resources available across a cluster of computers. Hadoop is inspired by a system invented at Google to create inverted index for its search product. Jeffrey Dean and Sanjay Ghemawat published papers in 2004 describing the system that they created for Google. The first one, titled "MapReduce: Simplified Data Processing on Large Clusters" is available at `research.google.com/archive/mapreduce.html`. The second one, titled "The Google File System" is available at `research.google.com/archive/gfs.html`. Inspired by these papers, Doug Cutting and Mike Cafarella developed an open source implementation, which later became Hadoop.

Many organizations have replaced expensive proprietary commercial products with Hadoop for processing large datasets. One reason is cost. Hadoop is open source and runs on a cluster of commodity hardware. You can scale it easily by adding cheap servers. High availability and fault tolerance are provided by Hadoop, so you don't need to buy expensive hardware. Second, it is better suited for certain types of data processing tasks, such as batch processing and ETL (extract transform load) of large-scale data.

Hadoop is built on a few important ideas. First, it is cheaper to use a cluster of commodity servers for both storing and processing large amounts of data than using high-end powerful servers. In other words, Hadoop uses scale-out architecture instead of scale-up architecture.

Second, implementing fault tolerance through software is cheaper than implementing it in hardware. Fault-tolerant servers are expensive. Hadoop does not rely on fault-tolerant servers. It assumes that servers will fail and transparently handles server failures. An application developer need not worry about handling hardware failures. Those messy details can be left for Hadoop to handle.

Third, moving code from one computer to another over a network is a lot more efficient and faster than moving a large dataset across the same network. For example, assume you have a cluster of 100 computers with a terabyte of data on each computer. One option for processing this data would be to move it to a very powerful server that can process 100 terabytes of data. However, moving 100 terabytes of data will take a long time, even on a very fast network. In addition, you will need very expensive hardware to process data with this approach. Another option is to move the code that processes this data to each computer in your 100-node cluster; it is a lot faster and more efficient than the first option. Moreover, you don't need high-end servers, which are expensive.

Fourth, writing a distributed application can be made easy by separating core data processing logic from distributed computing logic. Developing an application that takes advantage of resources available on a cluster of computers is a lot harder than developing an application that runs on a single computer. The pool of developers who can write applications that run on a single machine is several magnitudes larger than those who can write distributed applications. Hadoop provides a framework that hides the complexities of writing distributed applications. It thus allows organizations to tap into a much bigger pool of application developers.

Although people talk about Hadoop as a single product, it is not really a single product. It consists of three key components: a cluster manager, a distributed compute engine, and a distributed file system (see Figure 1-1).

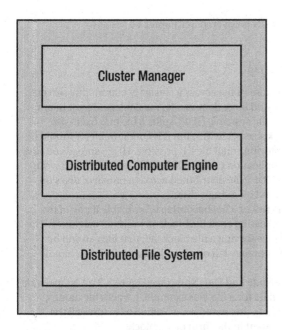

Figure 1-1. *Key conceptual Hadoop components*

Until version 2.0, Hadoop's architecture was monolithic. All the components were tightly coupled and bundled together. Starting with version 2.0, Hadoop adopted a modular architecture, which allows you to mix and match Hadoop components with non-Hadoop technologies.

The concrete implementations of the three conceptual components shown in Figure 1-1 are HDFS, MapReduce, and YARN (see Figure 1-2).

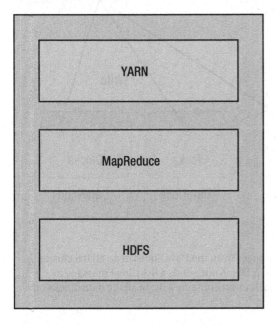

Figure 1-2. *Key Hadoop components*

HDFS and MapReduce are covered in this chapter. YARN is covered in Chapter 11.

HDFS (Hadoop Distributed File System)

HDFS, as the name implies, is a distributed file system. It stores a file across a cluster of commodity servers. It was designed to store and provide fast access to big files and large datasets. It is scalable and fault tolerant.

HDFS is a block-structured file system. Just like Linux file systems, HDFS splits a file into fixed-size blocks, also known as *partitions* or *splits*. The default block size is 128 MB, but it is configurable. It should be clear from the blocks' size that HDFS is not designed for storing small files. If possible, HDFS spreads out the blocks of a file across different machines. Therefore, an application can parallelize file-level read and write operations, making it much faster to read or write a large HDFS file distributed across a bunch of disks on different computers than reading or writing a large file stored on a single disk.

Distributing a file to multiple machines increases the risk of a file becoming unavailable if one of the machines in a cluster fails. HDFS mitigates this risk by replicating each file block on multiple machines. The default replication factor is 3. So even if one or two machines serving a file block fail, that file can still be read. HDFS was designed with the assumption that machines may fail on a regular basis. So it can handle failure of one or more machines in a cluster.

A HDFS cluster consists of two types of nodes: NameNode and DataNode (see Figure 1-3). A NameNode manages the file system namespace. It stores all the metadata for a file. For example, it tracks file names, permissions, and file block locations. To provide fast access to the metadata, a NameNode stores all the metadata in memory. A DataNode stores the actual file content in the form of file blocks.

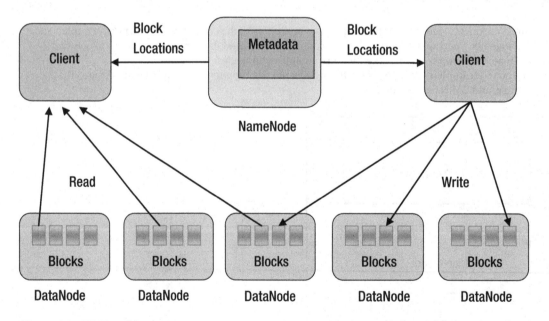

Figure 1-3. *HDFS architecture*

The NameNode periodically receives two types of messages from the DataNodes in an HDFS cluster. One is called Heartbeat and the other is called Blockreport. A DataNode sends a heartbeat message to inform the NameNode that it is functioning properly. A Blockreport contains a list of all the data blocks on a DataNode.

When a client application wants to read a file, it first contacts a NameNode. The NameNode responds with the locations of all the blocks that comprise that file. A block location identifies the DataNode that holds data for that file block. A client then directly sends a read request to the DataNodes for each file block. A NameNode is not involved in the actual data transfer from a DataNode to a client.

Similarly, when a client application wants to write data to an HDFS file, it first contacts the NameNode and asks it to create a new entry in the HDFS namespace. The NameNode checks whether a file with the same name already exists and whether the client has permissions to create a new file. Next, the client application asks the NameNode to choose DataNodes for the first block of the file. It creates a pipeline between all the replica nodes hosting that block and sends the data block to the first DataNode in the pipeline. The first DataNode stores the data block locally and forwards it to the second DataNode, which stores it locally and forwards it to the third DataNode. After the first file block has been stored on all the assigned DataNodes, the client asks the NameNode to select the DataNodes to host replicas of the second block. This process continues until all the file blocks have been stored on the DataNodes. Finally, the client informs the NameNode that the file writing is complete.

MapReduce

MapReduce is a distributed compute engine provided by Hadoop. While HDFS provides a distributed file system for storing large datasets, MapReduce provides a computing framework for processing large datasets in parallel across a cluster of computers. It abstracts cluster computing and provides higher-level constructs for writing distributed data processing applications. It enables programmers with no expertise in writing distributed or parallel applications to write applications that can run on a cluster of commodity computers.

The MapReduce framework automatically schedules an application's execution across a set of machines in a cluster. It handles load balancing, node failures, and complex internode communication. It takes care of the messy details of distributed computing and allows a programmer to focus on data processing logic.

The basic building blocks of a MapReduce application are two functions: map and reduce. Both primitives are borrowed from functional programming. All data processing jobs in a MapReduce application are expressed using these two functions. The map function takes as input a key-value pair and outputs a set of intermediate key-value pairs. The MapReduce framework calls the map function once for each key-value pair in the input dataset. Next, it sorts the output from the map functions and groups all intermediate values associated with the same intermediate key. It then passes them as input to the reduce function. The reduce function aggregates those values and outputs the aggregated value along with the intermediate key that it received as its input.

Spark, which is introduced in Chapter 3, is considered a successor to MapReduce. It provides many advantages over MapReduce. This is discussed in detail in Chapter 3.

Hive

Hive is data warehouse software that provides a SQL-like language for processing and analyzing data stored in HDFS and other Hadoop-compatible storage systems, such as Cassandra and Amazon S3. Although Hadoop made it easier to write data processing applications that can utilize the resources across a cluster of computers, the pool of programmers who can write such applications is still much smaller compared to the pool of people who know SQL.

SQL is one of the most widely used data processing languages. It is a declarative language. It looks deceptively simple, but it is a powerful language. SQL is easier to learn and use than Java and other programming languages used for writing a MapReduce application. Hive brought the simplicity of SQL to Hadoop and made it accessible to a wider user base.

Hive provides a SQL-like query language called Hive Query Language (HiveQL) for processing and analyzing data stored in any Hadoop-compatible storage system. It provides a mechanism to project a structure onto data stored in HDFS and query it using HiveQL. Under the hood, it translates HiveQL queries into MapReduce jobs. It also supports UDFs (user-defined functions) and UDAFs (user-defined aggregate functions), which can be used for complex data processing that cannot be efficiently expressed in HiveQL.

Spark SQL, which is discussed in Chapter 7, is considered a successor to Hive. However, Spark SQL provides more than just a SQL interface. It does a lot more, which is covered in detail in Chapter 7.

Data Serialization

Data has its own life cycle, independent of the program that creates or consumes it. Most of the time, data outlives the application that created it. Generally, it is saved on disk. Sometimes, it is sent from one application to another application over a network.

The format in which data is stored on disk or sent over a network is different from the format in which it lives in memory. The process of converting data in memory to a format in which it can be stored on disk or sent over a network is called *serialization*. The reverse process of reading data from disk or network into memory is called *deserialization*.

Data can be serialized using many different formats. Examples include CSV, XML, JSON, and various binary formats. Each format has pros and cons. For example, text formats such as CSV, XML, and JSON are human-readable, but not efficient in terms of either storage space or parse time. On the other hand, binary formats are more compact and can be parsed much quicker than text formats. However, binary formats are not human-readable.

The serialization/deserialization time or storage space difference between text and binary formats is not a big issue when a dataset is small. Therefore, people generally prefer text formats for small datasets, as they are easier to manage. However, for large datasets, the serialization/deserialization time or storage space difference between text and binary formats is significant. Therefore, binary formats are generally preferred for storing large datasets.

This section describes some of the commonly used binary formats for serializing big data.

Avro

Avro provides a compact language-independent binary format for data serialization. It can be used for storing data in a file or sending it over a network. It supports rich data structures, including nested data.

Avro uses a self-describing binary format. When data is serialized using Avro, schema is stored along with data. Therefore, an Avro file can be later read by any application. In addition, since schema is stored along with data, each datum is written without per-value overheads, making serialization fast and compact. When data is exchanged over a network using Avro, the sender and receiver exchange schemas during an initial connection handshake. An Avro schema is described using JSON.

Avro automatically handles field addition and removal, and forward and backward compatibility—all without any awareness by an application.

Thrift

Thrift is a language-independent data serialization framework. It primarily provides tools for serializing data exchange over a network between applications written in different programming languages. It supports a variety of languages, including C++, Java, Python, PHP, Ruby, Erlang, Perl, Haskell, C#, Cocoa, JavaScript, Node.js, Smalltalk, OCaml, Delphi, and other languages.

Thrift provides a code-generation tool and a set of libraries for serializing data and transmitting it across a network. It abstracts the mechanism for serializing data and transporting it across a network. Thus, it allows an application developer to focus on core application logic, rather than worry about how to serialize data and transmit it reliably and efficiently across a network.

With Thrift, an application developer defines data types and service interface in a language-neutral interface definition file. The services defined in an interface definition file is provided by a server application and used by a client application. The Thrift compiler compiles this file and generates code that a developer can then use to quickly build client and server applications.

A Thrift-based server and client can run on the same computer or different computers on a network. Similarly, the server and client application can be developed using the same programming language or different programming languages.

Protocol Buffers

Protocol Buffers is an open source data serialization framework developed by Google. Just like Thrift and Avro, it is language neutral. Google internally uses Protocol Buffers as its primary file format. It also uses it internally for exchanging data between applications over a network.

Protocol Buffers is similar to Thrift. It provides a compiler and a set of libraries that a developer can use to serialize data. A developer defines the structure or schema of a dataset in a file and compiles it with the Protocol Buffers compiler, which generates the code that can then be used to easily read or write that data.

Compared to Thrift, Protocol Buffers support a smaller set of languages. Currently, it supports C++, Java, and Python. In addition, unlike Thrift, which provides tools for both data serialization and building remote services, Protocol Buffers is primarily a data serialization format. It can be used for defining remote services, but it is not tied to any RPC (remote procedure call) protocol.

SequenceFile

SequenceFile is a binary flat file format for storing key-value pairs. It is commonly used in Hadoop as an input and output file format. MapReduce also uses SequenceFiles to store the temporary output from map functions.

A SequenceFile can have three different formats: Uncompressed, Record Compressed, and Block Compressed. In a Record Compressed SequenceFile, only the value in a record is compressed; whereas in a Block Compressed SequenceFile, both keys and values are compressed.

Columnar Storage

Data can be stored in either a row-oriented or a column-oriented format. In row-oriented formats, all columns or fields in a row are stored together. A row can be a row in a CSV file or a record in a database table. When data is saved using a row-oriented format, the first row is followed by the second row, which is followed by the third row, and so on. Row-oriented storage is ideal for applications that mostly perform CRUD (create, read, update, delete) operations on data. These applications operate on one row of data at a time.

However, row-oriented storage is not efficient for analytics applications. Such applications operate on the columns in a dataset. More importantly, these applications read and analyze only a small subset of columns across multiple rows. Therefore, reading all columns is a waste of memory, CPU cycles, and disk I/O, which is an expensive operation.

Another disadvantage of row-oriented storage is that data cannot be efficiently compressed. A record may consist of columns with different data types. Entropy is high across a row. Compression algorithms do not work very well on heterogeneous data. Therefore, a table stored on disk using row-oriented storage results in a larger file than that stored using columnar storage. A larger file not only consumes more disk space, but also impacts application performance, since disk I/O is proportional to file size and disk I/O is an expensive operation.

A column-oriented storage system stores data on disk by columns. All cells of a column are stored together or contiguously on disk. For example, when a table is saved on disk in a columnar format, data from all rows in the first column is saved first. It is followed by data from all rows in the second column, which is followed by a third column, and so on. Columnar storage is more efficient than row-oriented storage for analytic applications. It enables faster analytics and requires less disk space.

The next section discusses three commonly used columnar file formats in the Hadoop ecosystem.

RCFile

RCFile (Record Columnar File) was one of the first columnar storage formats implemented on top of HDFS for storing Hive tables. It implements a hybrid columnar storage format. RCFile first splits a table into row groups, and then stores each row group in columnar format. The row groups are distributed across a cluster.

RCFile allows you to take advantage of both columnar storage and Hadoop MapReduce. Since row groups are distributed across a cluster, they can be processed in parallel. Columnar storage of rows within a node allows efficient compression and faster analytics.

ORC

ORC (Optimized Row Columnar) is another columnar file format that provides a highly efficient way to store structured data. It provides many advantages over the RCFile format. For example, it stores row indexes, which allows it to quickly seek a given row during a query. It also provides better compression since it uses block-mode compression based on data type. Additionally, it can apply generic compression using zlib or Snappy on top of the data type based column-level compression.

Similar to RCFile, the ORC file format partitions a table into configurable-sized stripes (see Figure 1-4). The default stripe size is 250 MB. A stripe is similar to a row group in RCFile, but each stripe contains not only row data but also index data and a stripe footer. The stripe footer contains a directory of stream locations. Index data contains minimum and maximum values for each column, in addition to the row indexes. The ORC file format stores an index for every 10,000 rows in a stripe. Within each stripe, the ORC file format compresses columns using data type–specific encoding techniques such as run-length encoding for integer columns and dictionary encoding for string columns. It can further compress the columns using generic compression codecs, such as zlib or Snappy.

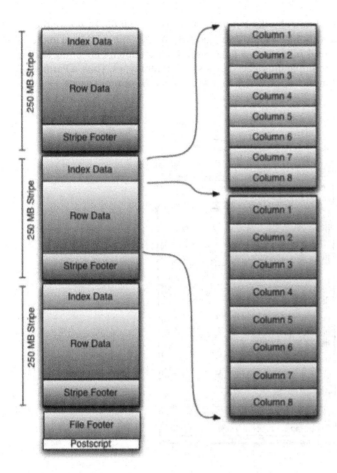

Figure 1-4. *ORC file structure (source: orc.apache.org)*

The stripes are followed by a file footer, which contains a list of stripes in a file, the number of rows in a stripe, and each columns data type. It also contains statistics for each column, such as count, min, max, and sum. The file footer is followed by a postscript section, which contains compression parameters and the size of the compressed footer.

The ORC file format not only stores data efficiently, but also allows efficient queries. An application can request only the columns needed in a query. Similarly, an application can skip reading entire set of rows using predicate pushdown.

Parquet

Parquet is yet another columnar storage format designed for the Hadoop ecosystem. It can be used with any data processing framework, including Hadoop MapReduce and Spark. It was designed to support complex nested data structures. In addition, it not only supports a variety of data encodings and compression techniques, but also allows compression schemes to be specified on a per-column basis.

Parquet implements a three-level hierarchical structure for storing data in a file (see Figure 1-5). First, it horizontally partitions a table into row groups, similar to RCFile and ORC. The row groups are distributed across a cluster and thus can be processed in parallel with any cluster-computing framework. Second, within each row group, it splits columns into column chunks. Parquet uses the term *column chunk* for the data in a column within a row group. A column chunk is stored contiguously on disk. The third level in the hierarchy is a *page*. Parquet splits a column chunk into pages. A page is the smallest unit for encoding and compression. A column chunk can consist of multiple interleaved pages of different types. Thus, a Parquet file consists of row groups, which contain column chunks, which in turn contain one or more pages.

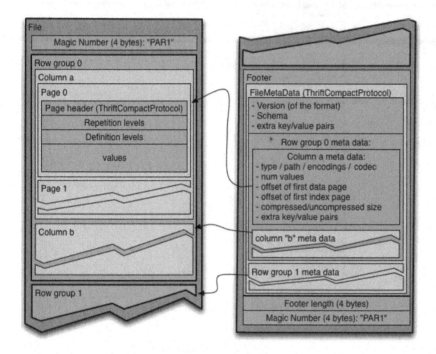

Figure 1-5. *Parquet file structure (source: parquet.apache.org)*

Messaging Systems

Data usually flows from one application to another. It is produced by one application and used by one or more other applications. Generally, the application generating or sending data is referred to as a *producer*, and the one receiving data is called a *consumer*.

Sometimes there is an asymmetry between the number of applications producing data and the number of applications consuming that data. For example, one application may produce data, which gets consumed by multiple consumers. Similarly, one application may consume data from multiple producers.

There is also sometimes asymmetry between the rate at which one application produces data and the rate at which another application can consume it. An application may produce data faster than the rate at which consumers can consume that data.

A simple way to send data from one application to another is to connect them to each other directly. However, this will not work if there is asymmetry either in the number of data producers and consumers or the rate at which data is produced and consumed. One more challenge is that tight coupling between producers and consumers requires them to be run at the same time or to implement a complex buffering mechanism. Therefore, direct connections between producers and consumers does not scale.

A flexible and scalable solution is to use a message broker or messaging system. Instead of applications connecting directly to each other, they connect to a message broker or a messaging system. This architecture makes it easy to add producers or consumers to a data pipeline. It also allows applications to produce and consume data at different rates.

This section discusses some of the messaging systems commonly used with big data applications.

Kafka

Kafka is a distributed messaging system or message broker. To be accurate, it is a distributed, partitioned, replicated commit log service, which can be used as a publish-subscribe messaging system.

Key features of Kafka include high throughput, scalability, and durability. A single broker can handle several hundred megabytes of reads and writes per second from thousands of applications. It can be easily scaled by adding more nodes to a cluster. For durability, it saves messages on disk.

The key entities in a Kafka-based architecture are brokers, producers, consumers, topics, and messages (see Figure 1-6). Kafka runs as a cluster of nodes, each of which is called a *broker*. Messages sent through Kafka are categorized into *topics*. An application that publishes messages to a Kafka topic is called a *producer*. A *consumer* is an application that subscribes to a Kafka topic and processes messages.

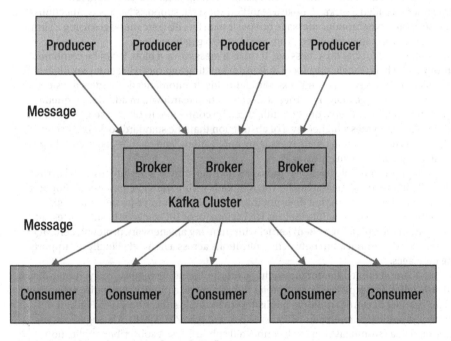

Figure 1-6. *Flow of messages through Kafka*

Kafka splits a topic into partitions. Each partition is an ordered immutable sequence of messages. New messages are appended to a partition. Each message in a partition is assigned a unique sequential identifier called *offset*. Partitions are distributed across the nodes in a Kafka cluster. In addition, they are replicated for fault tolerance. Partitioning of topics helps with scalability and parallelism. A topic need not fit on a single machine. It can grow to any size. Growth in topic size can be handled by adding more nodes to a Kafka cluster.

An important property of messages published to a Kafka cluster is that it retains all messages for a configurable period of time. Even after a consumer consumes a message, it is still available for the configured interval. More importantly, Kafka's performance is effectively constant with respect to data size.

Kafka supports both queuing and publish-subscribe messaging models using an abstraction called *consumer group*. Each message published to a topic is delivered to a single consumer within each subscribing consumer group. Thus, if all the consumers subscribing to a topic belong to the same consumer group, then Kafka acts as a queuing messaging system, where each message is delivered to only one consumer. On the other hand, if each consumer subscribing a topic belongs to a different consumer group, then Kafka acts as a publish-scribe messaging system, where each message is broadcast to all consumers subscribing a topic.

ZeroMQ

ZeroMQ is a lightweight high-performance messaging library. It is designed for implementing messaging queues and for building scalable concurrent and distributed message-driven applications. It does not impose a message broker-centric architecture, although it can be used to build a message broker if required. It supports most modern languages and operating systems.

The ZeroMQ API is modelled after the standard UNIX socket API. Applications communicate with each other using an abstraction called a *socket*. Unlike a standard socket, it supports N-to-N connections. A ZeroMQ socket represents an asynchronous message queue. It transfers discrete messages using a simple framing on the wire. Messages can range anywhere from zero bytes to gigabytes.

ZeroMQ does not impose any format on a message. It treats a message as a blob. It can be combined with a serialization protocol such as Google's protocol buffers for sending and receiving complex objects.

ZeroMQ implements I/O asynchronously in background threads. It automatically handles physical connection setup, reconnects, message delivery retries, and connection teardown. In addition, it queues messages if a recipient is unavailable. When a queue is full, it can be configured to block a sender or throw away messages. Thus, ZeroMQ provides a higher level of abstraction than the standard sockets for sending and receiving messages. It makes it easier to create messaging applications, and enables loose coupling between applications sending and receiving messages.

The ZeroMQ library supports multiple transport protocols for inter-thread, inter-process, and across the network messaging. For inter-thread messaging between threads within the same process, it supports a memory-based message passing transport that does not involve any I/O. For inter-process messaging between processes running on the same machine, it uses UNIX domain or IPC sockets. In such cases, all communication occurs within the operating system kernel without using any network protocol. ZeroMQ supports the TCP protocol for communication between applications across a network. Finally, it supports PGM for multicasting messages.

ZeroMQ can be used to implement different messaging patterns, including request-reply, router-dealer, client-server, publish-subscribe, and pipeline. For example, you can create a publish-subscribe messaging system with ZeroMQ for sending data from multiple publishers to multiple subscribers (see Figure 1-7). To implement this pattern, a publisher application creates a socket of type ZMQ_PUB. Messages sent on such sockets are distributed in a fan-out fashion to all connected subscribers. A subscriber application creates a socket of type ZMQ_SUB to subscribe to data published by a publisher. It can specify filters to specify messages of interest. Similarly, you can create a pipeline pattern with ZeroMQ to distribute data to nodes arranged in a pipeline. An application creates a socket of type ZMQ_PUSH to send messages to a downstream application, which creates a socket of type ZMQ_PULL.

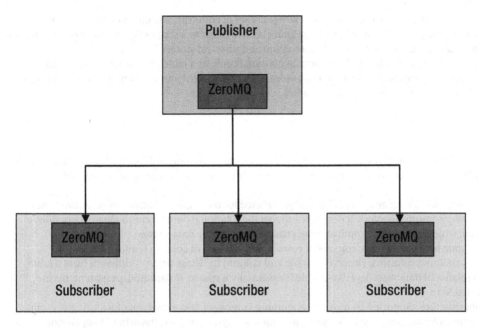

Figure 1-7. Publish-subscribe using ZeroMQ

NoSQL

The term NoSQL is used for a broad category of non-relational modern databases. Initially, NoSQL stood for "No SQL support" since these databases did not support SQL. However, now it means "Not only SQL," since some of these databases support a subset of SQL commands. The NoSQL databases have different design goals than RDBMS databases. A relational database guarantees ACID (Atomicity, Consistency, Isolation, Durability). A NoSQL database trades-off ACID compliance for linear scalability, performance, high-availability, flexible schema, and other features.

This section discusses some of the commonly used NoSQL databases.

Cassandra

Cassandra is a distributed, scalable, and fault-tolerant NoSQL database designed for storing large datasets. It is a partitioned row-store with tunable consistency. One of its key features is dynamic schema. Each row can store different columns, unlike relational databases where each row has the exact same columns. In addition, Cassandra is optimized for writes, so inserts are high-performant.

Cassandra has a masterless distributed architecture. Therefore, it does not have a single point of failure. In addition, it provides automatic distribution of rows across a cluster. A client application reading or writing data can connect to any node in a Cassandra cluster.

Cassandra provides high availability through built-in support for data replication. The number of replicas to be saved is configurable. Each replica gets stored on a different node in a cluster. So if the replication factor is 3, even if one or two nodes fail, the cluster is still available.

Data is modeled in Cassandra using a hierarchy of keyspace, table, row, and column. A keyspace is conceptually similar to a database or schema in an RDBMS. It is a logical collection of tables. It represents a namespace. It is designed for controlling data replication for a set of tables. A table, also known as a *column family*, is conceptually similar to a table in an RDBMS. A column family consists of a collection

13

of partitioned rows. Each row consists of a partition key and a set of columns. It is important to note that although a keyspace, table, row, and column in Cassandra seem similar to a schema, table, row, and column, respectively, in a relational database, their implementation and physical storage is different.

Query patterns drive data models in Cassandra. A column family or a table in Cassandra is basically a materialized view. Unlike relational databases, Cassandra does not support joins. This means the same data may need to be duplicated in multiple column families.

HBase

HBase is also a distributed, scalable, and fault-tolerant NoSQL data store designed for storing large datasets. It runs on top of HDFS. It has similar characteristics as Cassandra, since both are inspired by Bigtable, a data store invented by Google.

Bigtable is a distributed storage system that Google created for managing petabytes of structured data across thousands of commodity servers. It does not support the relational data model; instead, it provides a simple data model, which gives client applications dynamic control over data storage.

HBase stores data in tables. A table consists of rows. A row consists of column families. A column family consists of versioned columns. However, a table and column in HBase are very different from a table and column in a relational database. An HBase table is essentially a sparse, distributed, persistent, multi-dimensional, sorted Map.

Map is a data structure supported by most programming languages. It is a container for storing key-value pairs. It is a very efficient data structure for looking up values by keys. Generally, the order of keys is not defined and an application does not care about the order since it gives a key to the Map and gets back a value for that key. Note that the Map data structure should not be confused with the map function in Hadoop MapReduce. The map function is a functional language concept for transforming data.

The Map data structure is called by different names in different programming languages. For example, in PHP, it is called an *associative array*. In Python, it is known as a *dictionary*. Ruby calls it Hash. Java and Scala call it Map.

An HBase table is a sorted multi-dimensional or multi-level Map. The first level key is the row key, which allows an application to quickly read a row from billions of rows. The second level key is the column family. The third-level key is the column name, also known as a *column qualifier*. The fourth-level key is the timestamp. A combination of row key, column family, column, and timestamp uniquely identify a cell, which contains a value. A value is an un-interpreted array of bytes.

A row in an HBase table is sparse. Unlike rows in a relational database, not every row in HBase needs to have the same columns. Each row has the same set of column families, but a row may not store anything in some column families. An empty cell does not take any storage space.

Distributed SQL Query Engine

As discussed earlier, SQL is one of the most commonly used languages for querying and analyzing data. It is easy to learn and there are a lot more people who know SQL than those who know programming languages such as Java. Basically, Hive was created for this reason. However, Hive depends on MapReduce since it translates HiveQL queries into MapReduce jobs.

MapReduce is a powerful framework; however, it was designed for batch data processing. It has high throughput and high latency. It is great for data transformation or ETL (extract, transform, and load) jobs, but not an ideal platform for interactive queries or real-time analytics. Hive inherited the limitations of MapReduce. This motivated the creation of low-latency query engines using a different architecture.

This section discusses a few open source low-latency distributed SQL query engines that do not use MapReduce. Spark SQL can also act as a distributed query engine, but it is not covered here; it is discussed in detail in Chapter 7.

Impala

Impala is an open source data analytics software. It provides SQL interface for analyzing large datasets stored in HDFS and HBase. It supports HiveQL, the SQL-like language supported by Hive. It can be used for both batch and real-time queries.

Impala does not use MapReduce. Instead, it uses a specialized distributed query engine to avoid high latency. Its architecture is similar to commercially available MPP (massively parallel processing) databases. As a result, it generally provides an order-of-magnitude faster response time than Hive.

Presto

Presto is also an open source distributed SQL query engine for analyzing large datasets. Currently, it provides a SQL interface for analyzing data in HDFS, Cassandra, and relational databases. It allows interactive analytic queries on terabytes and petabytes of data. In addition, it allows a query to combine data from multiple sources.

Presto is architecturally similar to Impala. It does not use MapReduce for analyzing HDFS data; instead, it implements MPP architecture.

Apache Drill

Apache Drill is yet another open source distributed SQL query engine for analyzing large datasets stored in HDFS or NoSQL databases. It is inspired by Google's Dremel. It can be used to perform fast interactive ad hoc queries on petabytes of data. It implements a clustered MPP architecture, similar to Presto and Impala. It supports ANSI SQL and the JDBC/ODBC interface, so it can be used with any BI or data visualization application that supports JDBC/ODBC.

Key features of Drill include dynamic schema discovery, a flexible data model, decentralized metadata, and extensibility. Schema specification is not required to query a dataset with Drill. It uses information provided by self-describing formats such as Avro, JSON, Parquet, and NoSQL databases to determine the schema of a dataset. It can also handle schema changes during a query.

Drill supports a hierarchical data model that can be used to query complex data. It allows querying of complex nested data structures. For example, it can be used to query nested data stored in JSON or Parquet without the need to flatten them.

A centralized metadata store is not required with Drill. It gets metadata from the storage plug-in of a data source. Since it does not depend on a centralized metadata store, Drill can be used to query data from multiple sources, such Hive, HBase, and files at once. Thus, it can be used as a data virtualization platform.

Drill is compatible with Hive. It can be used in Hive environments to enable fast, interactive, ad hoc queries on existing Hive tables. It supports Hive metadata, UDFs (user-defined functions), and file formats.

Summary

Exponential growth in data in recent years has created opportunities for many big data technologies. The traditional proprietary products either cannot handle big data or are too expensive. This opened the door for open source big data technologies. Rapid innovation in this space has given rise to many new products, just in the last few years. The big data space has become so big that a book could be written to just introduce the various big data technologies.

Instead, this chapter discussed some of the big data technologies that get used along with Spark. It also introduced Hadoop and the key technologies in the Hadoop ecosystem. Spark is a part of this ecosystem too.

Spark is introduced in Chapter 3. Chapter 2 takes a detour and discusses Scala, which is a hybrid functional and object-oriented programming language. Understanding Scala is important, since all the code examples in this book are in Scala. In addition, Spark itself is written in Scala, although it supports other languages, including Java, Python, and R.

CHAPTER 2

■ ■ ■

Programming in Scala

Scala is one of the hottest modern programming languages. It is the Cadillac of programming languages. It is not only powerful but also a beautiful language. Learning Scala will provide a boost to your professional career.

Big data applications are written in a number of programming languages, including Java, Python, C++, Scala, and others. Hadoop itself is written in Java. Most Hadoop applications are written in Java, but it supports other programming languages too. Similarly, Spark is written in Scala, but it supports multiple programming languages, including Scala, Java Python, and R.

Scala is a great language for developing big data applications. It provides a number of benefits. First, a developer can achieve a significant productivity jump by using Scala. Second, it helps developers write robust code with reduced bugs. Third, Spark is written in Scala, so Scala is a natural fit for developing Spark applications.

This chapter introduces Scala as a general-purpose programming language. My goal is not to make you an expert in Scala, but to help you learn enough Scala to understand and write Spark applications in Scala. The sample code in this book is in Scala, so knowledge of Scala will make it easier to follow the material. If you already know Scala, you can safely skip this chapter.

With the preceding goal in mind, this chapter covers the fundamentals of programming in Scala. To effectively use Scala, it is important to known functional programming, so functional programming is introduced first. The chapter wraps up with a sample standalone Scala application.

Functional Programming (FP)

Functional programming is a programming style that uses functions as a building block and avoids mutable variables, loops, and other imperative control structures. It treats computation as an evaluation of mathematical functions, where the output of a function depends only on the arguments to the function. A program is composed of such functions. In addition, functions are first-class citizens in a functional programming language.

Functional programming has attracted a lot of attention in recent years. Even mainstream languages such as C++, Java, and Python have added support for functional programming. It has become popular for a few good reasons.

First, functional programming provides a tremendous boost in developer productivity. It enables you to solve a problem with fewer lines of code compared to imperative languages. For example, a task that requires 100 lines of code in Java may require only 10 or 20 lines of code in Scala. Thus, functional programming can increase your productivity five to ten times.

Second, functional programming makes it easier to write concurrent or multithreaded applications. The ability to write multi-threaded applications has become very important with the advent of multi-CPU or multi-core computers. Keeping up with Moore's law become harder and harder for hardware manufacturers, so instead of making processors faster, they started adding more CPUs and cores. Multi-core computers have become common today. Applications need to take advantage of all the cores. Functional programming languages make this task easier than imperative languages.

Third, functional programming helps you to write robust code. It helps you avoid common programming errors. In addition, the number of bugs in an application is generally proportional to the lines of code. Since functional programming requires a lot less lines of code compared to imperative programming, fewer bugs get into the code.

Finally, functional programming languages make it easier to write elegant code, which is easy to read, understand, and reason about. A properly written functional code looks beautiful; it is not complex or messy. You get immense joy and satisfaction from your code.

This section discusses key functional programming concepts.

Functions

A *function* is a block of executable code. Functions enable a programmer to split a large program into smaller manageable pieces. In functional programming, an application is built entirely by assembling functions.

Although many programming languages support the concept of functions, functional programming languages treat functions as a first-class citizen. In addition, in functional programming, functions are composable and do not have side effects.

First-Class

FP treats functions as first-class citizens. A function has the same status as a variable or value. It allows a function to be used just like a variable. It is easier to understand this concept if you contrast FP functions with functions in imperative languages such as C.

Imperative languages treat variables and functions differently. For example, C does not allow a function to be defined inside another function. It does not allow a function to be passed as an input parameter to another function.

FP allows a function to be passed as an input to another function. It allows a function to be returned as a return value from another function. A function can be defined anywhere, including inside another function. It can be defined as an unnamed function literal just like a string literal and passed as an input to a function.

Composable

Functions in functional programming are composable. Function composition is a mathematical and computer science concept of combining simple functions to create a complex one. For example, two composable functions can be combined to create a third function. Consider the following two mathematical functions:

$$f(x) = x*2$$
$$g(x) = x+2$$

The function f takes a numerical input and returns twice the value of the input as output. The function g also takes a numerical input and returns two plus that number as output.

A new function can be composed using f and g, as follows:

$$h(x) = f(g(x)) = f(x+2) = (x+2)*2$$

Using the function h is same as first calling function g with the input given to h, and then calling function f with the output from function g.

Function composability is a useful technique for solving a complex problem by breaking it into a bunch of simpler subproblems. Functions can then be written for each subproblem and assembled together in a top-level function.

No Side Effects

A function in functional programming does not have side effects. The result returned by a function depends only on the input arguments to the function. The behavior of a function does not change with time. It returns the same output every time for a given input, no matter how many times it is called. In other words, a function does not have a state. It does not depend on or update any global variable.

Functions with no side effects provide a number of benefits. First, they can be composed in any order. Second, it is easy to reason about the code. Third, it is easier to write multi-threaded applications with such functions.

Simple

Functions in functional programming are simple. A function consists of a few lines of code and it does only one thing. A simple function is easy to reason about. Thus, it enables robustness and high quality.

Simple composable functions make it easy to implement complex algorithms with confidence. FP encourages recursively splitting a problem into subproblems until you can solve a subproblem with a simple function. Then, the simple functions can be assembled to form a function that solves a complex problem.

Immutable Data Structures

Functional programming emphasizes the usage of immutable data structures. A purely functional program does not use any mutable data structure or variable. In other words, data is never modified in place, unlike in imperative programming languages such as C/C++, Java, and Python. People with no functional programming background find it difficult to imagine a program with no mutable variables. In practice, it is not hard to write code with immutable data structures.

Immutable data structures provide a number of benefits. First, they reduce bugs. It is easy to reason about code written with immutable data structures. In addition, functional languages provide constructs that allow a compiler to enforce immutability. Thus, many bugs are caught at compile time.

Second, immutable data structures make it easier to write multi-threaded applications. Writing an application that utilizes all the cores is not an easy task. Race conditions and data corruption are common problems with multi-threaded applications. Usage of immutable data structures helps avoid these problems.

Everything Is an Expression

In functional programming, every statement is an expression that returns a value. For example, the if-else control structure in Scala is an expression that returns a value. This behavior is different from imperative languages, where you can just group a bunch of statements within if-else.

This feature is useful for writing applications without mutable variables.

Scala Fundamentals

Scala is a hybrid programming language that supports both object-oriented and functional programming. It supports functional programming concepts such as immutable data structures and functions as first-class citizens. For object-oriented programming, it supports concepts such as class, object, and trait. It also supports encapsulation, inheritance, polymorphism, and other important object-oriented concepts.

Scala is a statically typed language. A Scala application is compiled by the Scala compiler. It is a type-safe language and the Scala compiler enforces type safety at compile time. This helps reduce the number of bugs in an application.

Finally, Scala is a Java virtual machine (JVM)–based language. The Scala compiler compiles a Scala application into Java bytecode, which will run on any JVM. At the bytecode level, a Scala application is indistinguishable from a Java application.

Since Scala is JVM-based, it is seamlessly interoperable with Java. A Scala library can be easily used from a Java application. More importantly, a Scala application can use any Java library without any wrapper or glue code. Thus, Scala applications benefit from the vast library of existing Java code that people have developed over the last two decades.

Although Scala is a hybrid object-oriented and functional programming language, it emphasizes functional programming. That is what makes it a powerful language. You will reap greater benefit from using Scala as a functional programming language than if you used it just as another object-oriented programming language.

Complete coverage of Scala is out-of-scope for this book. It would require a thick book to cover Scala in detail. Instead, only the fundamental constructs needed to write a Spark application are discussed. In addition, I assume that you have some programming experience, so I will not discuss the basics of programming.

Scala is a powerful language. With power comes complexity. Some people get intimidated with Scala because they try to learn all the language features at once. However, you do not need to know every bell and whistle in Scala to use it effectively. You can productively start developing Scala applications once you learn the fundamentals covered in this chapter.

Getting Started

The best way to learn a programming language is to program in it. You are able to better understand the material presented in this chapter if you play with the code samples as you read.

You can write Scala code in any text editor, compile it with scalac, and run it with scala. Alternatively, you can use the browser-based IDE provided by Typesafe. You can also use the Eclipsed-based Scala IDE, IntelliJ IDEA, or NetBeans IDE. You can download the Scala binaries and Typesafe Activator from www.scala-lang.org/download. The same site also provides links to download the Eclipsed-based Scala IDE, IntelliJ IDEA, or NetBeans IDE.

The easiest way to get started with Scala is by using the Scala interpreter, which provides an interactive shell for writing Scala code. It is a REPL (read, evaluate, print, loop) tool. When you type an expression in the Scala shell, it evaluates that expression, prints the result on the console, and waits for the next expression. Installing the interactive Scala shell is as simple as downloading the Scala binaries and unpackaging it. The Scala shell is called scala. It is located in the bin directory. You launch it by typing **scala** in a terminal.

```
$ cd /path/to/scala-binaries
$ bin/scala
```

At this point, you should see the Scala shell prompt, as shown in Figure 2-1.

```
Welcome to Scala version 2.11.7 (Java HotSpot(TM) 64-Bit Server VM, Java 1.7.0_67).
Type in expressions to have them evaluated.
Type :help for more information.

scala>
```

Figure 2-1. *The Scala shell prompt*

You can now type any Scala expression. An example is shown next.

```scala
scala> println("hello world")
```

After you press the Enter key, the Scala interpreter evaluates your code and prints the result on the console. You can use this shell for playing with the code samples shown in this chapter.

Let's begin learning Scala now.

Basic Types

Similar to other programming languages, Scala comes prepackaged with a list of basic types and operations allowed on those types. The list of basic types in Scala is shown in Table 2-1.

Table 2-1. *Basic Scala Variable Types*

Variable Type	Description
Byte	8-bit signed integer
Short	16-bit signed integer
Int	32-bit signed integer
Long	64-bit signed integer
Float	32-bit single precision float
Double	64-bit double precision float
Char	16-bit unsigned Unicode character
String	A sequence of Chars
Boolean	true or false

Note that Scala does not have primitive types. Each type in Scala is implemented as a class. When a Scala application is compiled to Java bytecode, the compiler automatically converts the Scala types to Java's primitive types wherever possible to optimize application performance.

Variables

Scala has two types of variables: mutable and immutable. Usage of mutable variables is highly discouraged. A pure functional program would never use a mutable variable. However, sometimes usage of mutable variables may result in less complex code, so Scala supports mutable variables too. It should be used with caution.

A mutable variable is declared using the keyword var; whereas an immutable variable is declared using the keyword val.

A var is similar to a variable in imperative languages such as C/C++ and Java. It can be reassigned after it has been created. The syntax for creating and modifying a variable is shown next.

```scala
var x = 10
x = 20
```

A val cannot be reassigned after it has been initialized. The syntax for creating a val is shown next.

```
val y = 10
```

What happens if later in the program, you add the following statement?

```
y = 20
```

The compiler will generate an error.

It is important to point out a few conveniences that the Scala compiler provides. First, semicolons at the end of a statement are optional. Second, the compiler infers type wherever possible. Scala is a statically typed language, so everything has a type. However, the Scala compiler does not force a developer to declare the type of something if it can infer it. Thus, coding in Scala requires less typing and the code looks less verbose.

The following two statements are equivalent.

```
val y: Int = 10;
val y = 10
```

Functions

As mentioned previously, a function is a block of executable code that returns a value. It is conceptually similar to a function in mathematics; it takes inputs and returns an output.

Scala treats functions as first-class citizens. A function can be used like a variable. It can be passed as an input to another function. It can be defined as an unnamed function literal, like a string literal. It can be assigned to a variable. It can be defined inside another function. It can be returned as an output from another function.

A function in Scala is defined with the keyword def. A function definition starts with the function name, which is followed by the comma-separated input parameters in parentheses along with their types. The closing parenthesis is followed by a colon, function output type, equal sign, and the function body in optional curly braces. An example is shown next.

```
def add(firstInput: Int, secondInput: Int): Int = {
  val sum = firstInput + secondInput
  return sum
}
```

In the preceding example, the name of the function is add. It takes two input parameters, both of which are of type Int. It returns a value, also of type Int. This function simply adds its two input parameters and returns the sum as output.

Scala allows a concise version of the same function, as shown next.

```
def add(firstInput: Int, secondInput: Int) = firstInput + secondInput
```

The second version does the exact same thing as the first version. The type of the returned data is omitted since the compiler can infer it from the code. However, it is recommended not to omit the return type of a function.

The curly braces are also omitted in this version. They are required only if a function body consists of more than one statement.

In addition, the keyword return is omitted since it is optional. Everything in Scala is an expression that returns a value. The result of the last expression, represented by the last statement, in a function body becomes the return value of that function.

The preceding code snippet is just one example of how Scala allows you to write concise code. It eliminates boilerplate code. Thus, it improves code readability and maintainability.

Scala supports different types of functions. Let's discuss them next.

Methods

A method is a function that is a member of an object. It is defined like and works the same as a function. The only difference is that a method has access to all the fields of the object to which it belongs.

Local Functions

A function defined inside another function or method is called a *local function*. It has access to the variables and input parameters of the enclosing function. A local function is visible only within the function in which it is defined. This is a useful feature that allows you to group statements within a function without polluting your application's namespace.

Higher-Order Methods

A method that takes a function as an input parameter is called a *higher-order method*. Similarly, a high-order function is a function that takes another function as input. Higher-order methods and functions help reduce code duplication. In addition, they help you write concise code.

The following example shows a simple higher-order function.

```scala
def encode(n: Int, f: (Int) => Long): Long = {
  val x = n * 10
  f(x)
}
```

The encode function takes two input parameters and returns a Long value. The first input type is an Int. The second input is a function f that takes an Int as input and returns a Long. The body of the encode function multiplies the first input by 10 and then calls the function that it received as an input.

You will see more examples of higher-order methods when Scala collections are discussed.

Function Literals

A function literal is an unnamed or anonymous function in source code. It can be used in an application just like a string literal. It can be passed as an input to a higher-order method or function. It can also be assigned to a variable.

A function literal is defined with input parameters in parenthesis, followed by a right arrow and the body of the function. The body of a functional literal is enclosed in optional curly braces. An example is shown next.

```scala
(x: Int) => {
          x + 100
        }
```

If the function body consists of a single statement, the curly braces can be omitted. A concise version of the same function literal is shown next.

```
(x: Int) => x + 100
```

The higher-order function encode defined earlier can be used with a function literal, as shown next.

```
val code = encode(10, (x: Int) => x + 100)
```

Closures

The body of a function literal typically uses only input parameters and local variables defined within the function literal. However, Scala allows a function literal to use a variable from its environment. A closure is a function literal that uses a non-local non-parameter variable captured from its environment. Sometimes people use the terms *function literal* and *closure* interchangeably, but technically, they are not the same.

The following code shows an example of a closure.

```
def encodeWithSeed(num: Int, seed: Int): Long = {

  def encode(x: Int, func: (Int) => Int): Long = {
    val y = x + 1000
    func(y)
  }

  val result = encode(num, (n: Int) => (n * seed))
  result
}
```

In the preceding code, the local function encode takes a function as its second parameter. The function literal passed to encode uses two variables n and seed. The variable n was passed to it as a parameter; however, seed is not passed as parameter. The function literal passed to the encode function captures the variable seed from its environment and uses it.

Classes

A class is an object-oriented programming concept. It provides a higher-level programming abstraction. At a very basic level, it is a code organization technique that allows you to bundle data and all of its operations together. Conceptually, it represents an entity with properties and behavior.

A class in Scala is similar to that in other object-oriented languages. It consists of fields and methods. A field is a variable, which is used to store data. A method contains executable code. It is a function defined inside a class. A method has access to all the fields of a class.

A class is a template or blueprint for creating objects at runtime. An object is an instance of a class. A class is defined in source code, whereas an object exists at runtime. A class is defined using the keyword class. A class definition starts with the class name, followed by comma-separated class parameters in parentheses, and then fields and methods enclosed in curly braces.

An example is shown next.

```
class Car(mk: String, ml: String, cr: String) {
  val make = mk
  val model = ml
  var color = cr
```

```scala
  def repaint(newColor: String) = {
    color = newColor
  }
}
```

An instance of a class is created using the keyword new.

```scala
val mustang = new Car("Ford", "Mustang", "Red")
val corvette = new Car("GM", "Corvette", "Black")
```

A class is generally used as a mutable data structure. An object has a state, which changes with time. Therefore, a class may have fields that are defined using var.

Since Scala runs on the JVM, you do not need to explicitly delete an object. The Java garbage collector automatically removes objects that are no longer in use.

Singletons

One of the common design patterns in object-oriented programming is to define a class that can be instantiated only once. A class that can be instantiated only once is called a *singleton*. Scala provides the keyword object for defining a singleton class.

```scala
object DatabaseConnection {
  def open(name: String): Int = {
    ...
  }
  def read (streamId: Int): Array[Byte] = {
    ...
  }
  def close (): Unit = {
    ...
  }
}
```

Case Classes

A case class is a class with a case modifier. An example is shown next.

```scala
case class Message(from: String, to: String, content: String)
```

Scala provides a few syntactic conveniences to a case class. First, it creates a factory method with the same name. Therefore, you can create an instance of a case class without using the keyword new. For example, the following code is valid.

```scala
val request = Message("harry", "sam", "fight")
```

Second, all input parameters specified in the definition of a case class implicitly get a val prefix. In other words, Scala treats the case class Message as if it was defined, as shown next.

```scala
class Message(val from: String, val to: String, val content: String)
```

The val prefix in front of a class parameter turns it into a non-mutable class field. It becomes accessible from outside the class.

Third, Scala adds these four methods to a case class: toString, hashCode, equals, and copy. These methods make it easy to use a case class.

Case classes are useful for creating non-mutable objects. In addition, they support pattern matching, which is described next.

Pattern Matching

Pattern matching is a Scala concept that looks similar to a switch statement in other languages. However, it is a more powerful tool than a switch statement. It is like a Swiss Army knife that can be used for solving a number of different problems.

One simple use of pattern matching is as a replacement for a multi-level if-else statement. An if-else statement with more than two branches becomes harder to read. Use of pattern matching in such situations improves code readability.

As an example, consider a simple function that takes as an input parameter a string representing a color and returns 1 if the input string is "Red", 2 for "Blue", 3 for "Green", 4 for "Yellow", and 0 for any other color.

```scala
def colorToNumber(color: String): Int => {
  val num = color match {
              case "Red" => 1
              case "Blue" => 2
              case "Green" => 3
              case "Yellow" => 4
              case _ => 0
            }
  num
}
```

Instead of the keyword switch, Scala uses the keyword match. Each possible match is preceded by the keyword case. If there is a match for a case, the code on the right-hand side of the right arrow is executed. The underscore represents the default case. If none of the prior cases match, the code for the default case is executed.

The function shown earlier is a simple example, but it helps illustrate a few important things about pattern matching. First, once a match is found, only the code for the matched case is executed. Unlike a switch statement, a break statement is not required after the code for each case. Code execution does not fall through the remaining cases.

Second, the code on the right-hand side of each right arrow is an expression returning a value. Therefore, a pattern-matching statement itself is an expression returning a value. The following example illustrates this point better.

```scala
def f(x: Int, y: Int, operator: String): Double = {
  operator match {
    case "+" => x + y
    case "-" => x - y
    case "*" => x * y
    case "/" => x / y.toDouble
  }
}

val sum = f(10,20, "+")
val product = f(10, 20, "*")
```

Operators

Scala provides a rich set of operators for the basic types. However, it does not have built-in operators. In Scala, every type is a class and every operator is a method. Using an operator is equivalent to calling a method. Consider the following example.

```
val x = 10
val y = 20
val z = x + y
```

+ is not a built-in operator in Scala. It is a method defined in class Int. The last statement in the preceding code is same as the following code.

```
val z = x.+(y)
```

Scala allows you to call any method using the operator notation.

Traits

A *trait* represents an interface supported by a hierarchy of related classes. It is an abstraction mechanism that helps development of modular, reusable, and extensible code.

Conceptually, an interface is defined by a set of methods. An interface in Java only includes method signatures. Every class that inherits an interface must provide an implementation of the interface methods.

Scala traits are similar to Java interfaces. However, unlike a Java interface, a Scala trait can include implementation of a method. In addition, a Scala trait can include fields. A class can reuse the fields and methods implemented in a trait.

A trait looks similar to an abstract class. Both can contain fields and methods. The key difference is that a class can inherit from only one class, but it can inherit from any number of traits.

An example of a trait is shown next.

```
trait Shape {
    def area(): Int
}
class Square(length: Int) extends Shape {
    def area = length * length
}
class Rectangle(length: Int, width: Int) extends Shape {
    def area = length * width
}
val square = new Square(10)
val area = square.area
```

Tuples

A *tuple* is a container for storing two or more elements of different types. It is immutable; it cannot be modified after it has been created. It has a lightweight syntax, as shown next.

```
val twoElements = ("10", true)
val threeElements =  ("10", "harry", true)
```

A tuple is useful in situations where you want to group non-related elements. If the elements are of the same type, you can use a collection, such as an array or a list. If the elements are of different types, but related, you can store them as fields in a class. However, a class may be overkill in some situations. For example, you may have a function that returns more than one value. A tuple may be more appropriate in such cases.

An element in a tuple has a one-based index. The following code sample shows the syntax for accessing elements in a tuple.

```
val first = threeElements._1
val second = threeElements._2
val third = threeElements._3
```

Option Type

An Option is a data type that indicates the presence or absence of some data. It represents optional values. It can be an instance of either a case class called Some or singleton object called None. An instance of Some can store data of any type. The None object represents absence of data.

The Option data type is used with a function or method that optionally returns a value. It returns either Some(x), where x is the actual returned value, or the None object, which represents a missing value. The optional value returned by a function can be read using pattern matching.

The following code shows sample usage.

```
def colorCode(color: String): Option[Int] = {
  color match {
    case "red" => Some(1)
    case "blue" => Some(2)
    case "green" => Some(3)
    case _ => None
  }
}

val code = colorCode("orange")
code match {
  case Some(c) => println("code for orange is: " + c)
  case None => println("code not defined for orange")
}
```

The Option data type helps prevent null pointer exceptions. In many languages, null is used to represent absence of data. For example, a function in C/C++ or Java may be defined to return an integer. However, a programmer may return null if no valid integer can be returned for a given input. If the caller does not check for null and blindly uses the returned value, the program will crash. The combination of strong type checking and the Option type in Scala prevent these kinds of errors.

Collections

A collection is a container data structure. It contains zero or more elements. Collections provide a higher-level abstraction for working with data. They enable declarative programming. With an easy-to-use interface, they eliminate the need to manually iterate or loop through all the elements.

Scala has a rich collections library that includes collections of many different types. In addition, all the collections expose the same interface. As a result, once you become familiar with one Scala collection, you can easily use other collection types.

Scala collections can be broadly grouped into three categories: sequences, sets, and maps. This section introduces the commonly used Scala collections.

Sequences

A *sequence* represents a collection of elements in a specific order. Since the elements have a defined order, they can be accessed by their position in a collection. For example, you can ask for the *nth* element in a sequence.

Array

An *Array* is an indexed sequence of elements. All the elements in an array are of the same type. It is a mutable data structure; you can update an element in an array. However, you cannot add an element to an array after it has been created. It has a fixed length.

A Scala array is similar to an array in other languages. You can efficiently access any element in an array in constant time. Elements in an array have a zero-based index. To get or update an element, you specify its index in parenthesis. An example is shown next.

```
val arr = Array(10, 20, 30, 40)
arr(0) = 50
val first = arr(0)
```

Basic operations on an array include

- Fetching an element by its index

- Updating an element using its index

List

A *List* is a linear sequence of elements of the same type. It is a recursive data structure, unlike an array, which is a flat data structure. In addition, unlike an array, it is an immutable data structure; it cannot be modified after it has been created. List is one of the most commonly used data structures in Scala and other functional languages.

Although an element in a list can be accessed by its index, it is not an efficient data structure for accessing elements by their indices. Access time is proportional to the position of an element in a list.

The following code shows a few ways to create a list.

```
val xs = List(10,20,30,40)
val ys = (1 to 100).toList
val zs = someArray.toList
```

Basic operations on a list include

- Fetching the first element. For this operation, the List class provides a method named head.

- Fetching all the elements except the first element. For this operation, the List class provides a method named tail.

- Checking whether a list is empty. For this operation, the List class provides a method named isEmpty, which returns true if a list is empty.

Vector

The Vector class is a hybrid of the List and Array classes. It combines the performance characteristics of both Array and List. It provides constant-time indexed access and constant-time linear access. It allows both fast random access and fast functional updates.

An example is shown next.

```
val v1 = Vector(0, 10, 20, 30, 40)
val v2 = v1 :+ 50
val v3 = v2 :+ 60
val v4 = v3(4)
val v5 = v3(5)
```

Sets

Set is an unordered collection of distinct elements. It does not contain duplicates. In addition, you cannot access an element by its index, since it does not have one.

An example of a set is shown next.

```
val fruits = Set("apple", "orange", "pear", "banana")
```

Sets support two basic operations.

- contains: Returns true if a set contains the element passed as input to this method.

- isEmpty: Returns true if a set is empty.

Map

Map is a collection of key-value pairs. In other languages, it known as a dictionary, associative array, or hash map. It is an efficient data structure for looking up a value by its key. It should not be confused with the map in Hadoop MapReduce. That map refers to an operation on a collection.

The following code snippet shows how to create and use a Map.

```
val capitals = Map("USA" -> "Washington D.C.", "UK" -> "London", "India" -> "New Delhi")
val indiaCapital = capitals("India")
```

Scala supports a large number of collection types. Covering all of them is out of the scope for this book. However, a good understanding of the ones covered in this section will be enough to start productively using Scala.

Higher-Order Methods on Collection Classes

The real power of Scala collections comes from their higher-order methods. A higher-order method takes a function as its input parameter. It is important to note that a higher-order method does not mutate a collection.

This section discusses some of the most commonly used higher methods. The List collection is used in the examples, but all Scala collections support these higher-order methods.

map

The map method of a Scala collection applies its input function to all the elements in the collection and returns another collection. The returned collection has the exact same number of elements as the collection on which map was called. However, the elements in the returned collection need not be of the same type as that in the original collection.

An example is shown next.

```scala
val xs = List(1, 2, 3, 4)
val ys = xs.map((x: Int) => x * 10.0)
```

It should be noted that in the preceding example, xs is of type List[Int]; whereas ys is of type List[Double].

If a function takes a single argument, opening and closing parentheses can be replaced with opening and closing curly braces, respectively. The two statements shown next are equivalent.

```scala
val ys = xs.map((x: Int) => x * 10.0)
val ys = xs.map{(x: Int) => x * 10.0}
```

As mentioned earlier in this chapter, Scala allows you to call any method using operator notation. To further improve readability, the preceding code can also be written as follows.

```scala
val ys = xs map {(x: Int) => x * 10.0}
```

Scala can infer the type of the parameter passed to a function literal from the type of a collection, so you can omit the parameter type. The two following two statements are equivalent.

```scala
val ys = xs map {(x: Int) => x * 10.0}
val ys = xs map {x => x * 10.0}
```

If an input to a function literal is used only once in its body, the right arrow and left-hand side of the right arrow can be dropped from a function literal. You can write just the body of the function literal. The following two statements are equivalent.

```scala
val ys = xs map {x => x * 10.0}
val ys = xs map {_ * 10.0}
```

The underscore character represents an input to the function literal passed to the map method. The preceding code can be read as multiplying each element in the collection xs by 10.

To summarize, the following code sample shows both the verbose and concise version of the same statement.

```scala
val ys = xs.map((x: Int) => x * 10.0)
val ys = xs map {_ * 10.0}
```

As you can see, Scala makes it easier to write concise code, which is sometimes also easier to read.

flatMap

The flatMap method of a Scala collection is similar to map. It takes a function as input, applies it to each element in a collection, and returns another collection as a result. However, the function passed to flatMap generates a collection for each element in the original collection. Thus, the result of applying the input function is a collection of collections. If the same input function were passed to the map method, it would return a collection of collections. The flatMap method instead returns a flattened collection.

The following example illustrates a use of flatMap.

```
val line = "Scala is fun"
val SingleSpace = " "
val words = line.split(SingleSpace)
val arrayOfChars = words flatMap {_.toList}
```

The toList method of a collection creates a list of all the elements in the collection. It is a useful method for converting a string, an array, a set, or any other collection type to a list.

filter

The filter method applies a predicate to each element in a collection and returns another collection consisting of only those elements for which the predicate returned true. A predicate is function that returns a Boolean value. It returns either true or false.

```
val xs = (1 to 100).toList
val even = xs filter {_ %2 == 0}
```

foreach

The foreach method of a Scala collection calls its input function on each element of the collection, but does not return anything. It is similar to the map method. The only difference between the two methods is that map returns a collection and foreach does not return anything. It is a rare method that is used for its side effects.

```
val words = "Scala is fun".split(" ")
words.foreach(println)
```

reduce

The reduce method returns a single value. As the name implies, it reduces a collection to a single value. The input function to the reduce method takes two inputs at a time and returns one value. Essentially, the input function is a binary operator that must be both associative and commutative.

The following code shows some examples.

```
val xs = List(2, 4, 6, 8, 10)
val sum    = xs reduce {(x,y) => x + y}
val product  = xs reduce {(x,y) => x * y}
val max = xs reduce {(x,y) => if (x > y) x else y}
val min = xs reduce {(x,y) => if (x < y) x else y}
```

Here is another example that finds the longest word in a sentence

```scala
Val words = "Scala is fun" split(" ")
val longestWord = words reduce {(w1, w2) => if(w1.length > w2.length) w1 else w2}
```

Note that the map and reduce operations in Hadoop MapReduce are similar to the map and reduce methods that we discussed in this section. In fact, Hadoop MapReduce borrowed these concepts from functional programming.

A Standalone Scala Application

So far, you have seen just snippets of Scala code. In this section, you will write a simple yet complete standalone Scala application that you can compile and run.

A standalone Scala application needs to have a singleton object with a method called main. This main method takes an input of type Array[String] and does not return any value. It is the entry point of a Scala application. The singleton object containing the main method can be named anything.

A Scala application that prints "Hello World!" is shown next.

```scala
object HelloWorld {
    def main(args: Array[String]): Unit = {
      println("Hello World!")
    }
}
```

You can put the preceding code in a file, and compile and run it. Scala source code files have the extension .scala. It is not required, but recommended to name a Scala source file after the class or object defined in that file. For example, you would put the preceding code in a file named HelloWorld.scala.

Summary

Scala is a powerful JVM-based statically typed language for developing multi-threaded and distributed applications. It combines the best of both object-oriented and functional programming. In addition, it is seamlessly interoperable with Java. You can use any Java library from Scala and vice-versa.

The key benefits of using Scala include a significant jump in developer productivity and code quality. In addition, it makes it easier to develop robust multi-threaded and distributed applications.

Spark is written in Scala. It is just one example of the many popular distributed systems built with Scala.

CHAPTER 3

Spark Core

Spark is the most active open source project in the big data world. It has become hotter than Hadoop. It is considered the successor to Hadoop MapReduce, which we discussed in Chapter 1. Spark adoption is growing rapidly. Many organizations are replacing MapReduce with Spark.

Conceptually, Spark looks similar to Hadoop MapReduce; both are designed for processing big data. They both enable cost-effective data processing at scale using commodity hardware. However, Spark offers many advantages over Hadoop MapReduce. These are discussed in detail later in this chapter.

This chapter covers Spark core, which forms the foundation of the Spark ecosystem. It starts with an overview of Spark core, followed by the high-level architecture and runtime view of an application running on Spark. The chapter also discusses Spark core's programming interface.

Overview

Spark is an in-memory cluster computing framework for processing and analyzing large amounts of data. It provides a simple programming interface, which enables an application developer to easily use the CPU, memory, and storage resources across a cluster of servers for processing large datasets.

Key Features

The key features of Spark include the following:

- Easy to use
- Fast
- General-purpose
- Scalable
- Fault tolerant

Easy to Use

Spark provides a simpler programming model than that provided by MapReduce. Developing a distributed data processing application with Spark is a lot easier than developing the same application with MapReduce.

Spark offers a rich application programming interface (API) for developing big data applications; it comes with 80-plus data processing operators. Thus, Spark provides a more expressive API than that offered by Hadoop MapReduce, which provides just two operators: map and reduce. Hadoop MapReduce requires every problem to be broken down into a sequence of map and reduce jobs. It is hard to express non-trivial algorithms with just map and reduce. The operators provided by Spark make it easier to do complex data processing in Spark than in Hadoop MapReduce.

In addition, Spark enables you to write more concise code compared to Hadoop MapReduce, which requires a lot of boilerplate code. A data processing algorithm that requires 50 lines of code in Hadoop MapReduce can be implemented in less than 10 lines of code in Spark. The combination of a rich expressive API and the elimination of boilerplate code significantly increases developer productivity. A developer can be five to ten times more productive with Spark as compared to MapReduce.

Fast

Spark is orders of magnitude faster than Hadoop MapReduce. It can be hundreds of times faster than Hadoop MapReduce if data fits in memory. Even if data does not fit in memory, Spark can be up to ten times faster than Hadoop MapReduce.

Speed is important, especially when processing large datasets. If a data processing job takes days or hours, it slows down decision-making. It reduces the value of data. If the same processing can be run 10 to 100 times faster, it opens the door to many new opportunities. It becomes possible to develop new data-driven applications that were not possible before.

Spark is faster than Hadoop MapReduce for two reasons. First, it allows in-memory cluster computing. Second, it implements an advanced execution engine.

Spark's in-memory cluster computing capabilities provides an orders of magnitude performance boost. The sequential read throughput when reading data from memory compared to reading data from a hard disk is 100 times greater. In other words, data can be read from memory 100 times faster than from disk. The difference in read speed between disk and memory may not be noticeable when an application reads and processes a small dataset. However, when an application reads and processes terabytes of data, I/O latency (the time it takes to load data from disk to memory) becomes a significant contributor to overall job execution time.

Spark allows an application to cache data in memory for processing. This enables an application to minimize disk I/O. A MapReduce-based data processing pipeline may consist of a sequence of jobs, where each job reads data from disk, processes it, and writes the results to disk. Thus, a complex data processing application implemented with MapReduce may read data from and write data to disk several times. Since Spark allows caching of data in memory, the same application implemented with Spark reads data from disk only once. Once data is cached in memory, each subsequent operation can be performed directly on the cached data. Thus, Spark enables an application to minimize I/O latency, which, as previously mentioned, can be a significant contributor to overall job execution time.

Note that Spark does not automatically cache input data in memory. A common misconception is that Spark cannot be used if input data does not fit in memory. It is not true. Spark can process terabytes of data on a cluster that may have only 100 GB total cluster memory. It is up to an application to decide what data should be cached and at what point in a data processing pipeline that data should be cached. In fact, if a data processing application makes only a single pass over data, it need not cache data at all.

The second reason Spark is faster than Hadoop MapReduce is that it has an advanced job execution engine. Both Spark and MapReduce convert a job into a directed acyclic graph (DAG) of stages. In case you are not familiar with graph theory, a *graph* is a collection of vertices connected by edges. A *directed graph* is a graph with edges that have a direction. An *acyclic graph* is a graph with no graph cycles. Thus, a DAG is a directed graph with no directed cycles. In other words, there is no way you can start at some vertex in a DAG and follow a sequence of directed edges to get back to the same vertex. Chapter 11 provides a more detailed introduction to graphs.

Hadoop MapReduce creates a DAG with exactly two predefined stages—Map and Reduce—for every job. A complex data processing algorithm implemented with MapReduce may need to be split into multiple jobs, which are executed in sequence. This design prevents Hadoop MapReduce from doing any optimization.

In contrast, Spark does not force a developer to split a complex data processing algorithm into multiple jobs. A DAG in Spark can contain any number of stages. A simple job may have just one stage, whereas a complex job may consist of several stages. This allows Spark to do optimizations that are not possible with MapReduce. Spark executes a multi-stage complex job in a single run. Since it has knowledge of all the stages, it optimizes them. For example, it minimizes disk I/O and data shuffles, which involves data movement across a network and increases application execution time.

General Purpose

Spark provides a unified integrated platform for different types of data processing jobs. It can be used for batch processing, interactive analysis, stream processing, machine learning, and graph computing. In contrast, Hadoop MapReduce is designed just for batch processing. Therefore, a developer using MapReduce has to use different frameworks for stream processing and graph computing.

Using different frameworks for different types of data processing jobs creates many challenges. First, a developer has to learn multiple frameworks, each of which has a different interface. This reduces developer productivity. Second, each framework may operate in a silo. Therefore, data may have to be copied to multiple places. Similarly, code may need to be duplicated at multiple places. For example, if you want to process historical data with MapReduce and streaming data with Storm (a stream-processing framework) exactly the same way, you have to maintain two copies of the same code—one in Hadoop MapReduce and the other in Storm. The third problem with using multiple frameworks is that it creates operational headaches. You need to set up and manage a separate cluster for each framework. It is more difficult to manage multiple clusters than a single cluster.

Spark comes pre-packaged with an integrated set of libraries for batch processing, interactive analysis, stream processing, machine learning, and graph computing. With Spark, you can use a single framework to build a data processing pipeline that involves different types of data processing tasks. There is no need to learn multiple frameworks or deploy separate clusters for different types of data processing jobs. Thus, Spark helps reduce operational complexity and avoids code as well as data duplication.

Interestingly, many popular applications and libraries that initially used MapReduce as an execution engine are either being migrated to or adding support for Spark. For example, Apache Mahout, a machine-learning library initially built on top of Hadoop MapReduce, is migrating to Spark. In April 2014, the Mahout developers said goodbye to MapReduce and stopped accepting new MapReduce-based machine learning algorithms.

Similarly, the developers of Hive (discussed in Chapter 1) are developing a version that runs on Spark. Pig, which provides a scripting language for building data processing pipelines, has also added support for Spark as an execution engine. Cascading, an application development platform for building data applications on Hadoop, is also adding support for Spark.

Scalable

Spark is scalable. The data processing capacity of a Spark cluster can be increased by just adding more nodes to a cluster. You can start with a small cluster, and as your dataset grows, you can add more computing capacity. Thus, Spark allows you to scale economically.

In addition, Spark makes this feature automatically available to an application. No code change is required when you add a node to a Spark cluster.

Fault Tolerant

Spark is fault tolerant. In a cluster of a few hundred nodes, the probability of a node failing on any given day is high. The hard disk may crash or some other hardware problem may make a node unusable. Spark automatically handles the failure of a node in a cluster. Failure of a node may degrade performance, but will not crash an application.

Since Spark automatically handles node failures, an application developer does not have to handle such failures in his application. It simplifies application code.

Ideal Applications

As discussed in the previous section, Spark is a general-purpose framework; it can be used for a variety of big data applications. However, it is ideal for big data applications where speed is very important. Two examples of such applications are applications that allow interactive analysis and applications that use iterative data processing algorithms.

Iterative Algorithms

Iterative algorithms are data processing algorithms that iterate over the same data multiple times. Applications that use iterative algorithms include machine learning and graph processing applications. These applications run tens or hundreds of iterations of some algorithm over the same data. Spark is ideal for such applications.

The reason iterative algorithms run fast on Spark is its in-memory computing capabilities. Since Spark allows an application to cache data in memory, an iterative algorithm, even if it runs 100 iterations, needs to read data from disk only for the first iteration. Subsequent iterations read data from memory. Since reading data from memory is 100 times faster than reading from disk, such applications run orders of magnitude faster on Spark.

Interactive Analysis

Interactive data analysis involves exploring a dataset interactively. For example, it is useful to do a summary analysis on a very large dataset before firing up a long-running batch processing job that may run for hours. Similarly, a business analyst may want to interactively analyze data using a BI or data visualization tool. In such cases, a user runs multiple queries on the same data. Spark provides an ideal platform for interactively analyzing a large dataset.

The reason Spark is ideal for interactive analysis is, again, its in-memory computing capabilities. An application can cache the data that will be interactively analyzed in memory. The first query reads data from disk, but subsequent queries read the cached data from memory. Queries on data in memory execute orders of magnitude faster than on data on disk. A query that takes more than an hour when it reads data from disk may take seconds when it is run on the same data cached in memory.

High-level Architecture

A Spark application involves five key entities: a driver program, a cluster manager, workers, executors, and tasks (see Figure 3-1).

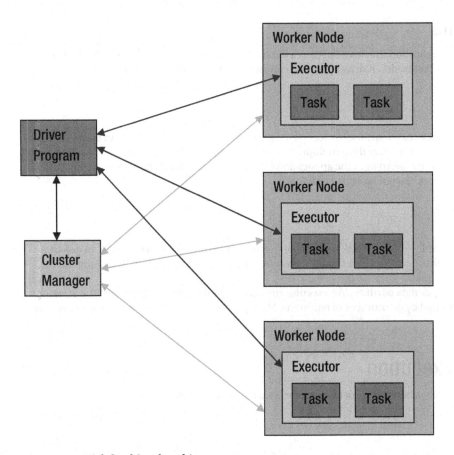

Figure 3-1. *High-level Spark architecture*

Workers

A worker provides CPU, memory, and storage resources to a Spark application. The workers run a Spark application as distributed processes on a cluster of nodes.

Cluster Managers

Spark uses a cluster manager to acquire cluster resources for executing a job. A cluster manager, as the name implies, manages computing resources across a cluster of worker nodes. It provides low-level scheduling of cluster resources across applications. It enables multiple applications to share cluster resources and run on the same worker nodes.

Spark currently supports three cluster managers: standalone, Mesos, and YARN. Mesos and YARN allow you to run Spark and Hadoop applications simultaneously on the same worker nodes. These cluster managers are discussed in more detail in Chapter 10.

Driver Programs

A driver program is an application that uses Spark as a library. It provides the data processing code that Spark executes on the worker nodes. A driver program can launch one or more jobs on a Spark cluster.

Executors

An executor is a JVM (Java virtual machine) process that Spark creates on each worker for an application. It executes application code concurrently in multiple threads. It can also cache data in memory or disk.

An executor has the same lifespan as the application for which it is created. When a Spark application terminates, all the executors created for it also terminate.

Tasks

A task is the smallest unit of work that Spark sends to an executor. It is executed by a thread in an executor on a worker node. Each task performs some computations to either return a result to a driver program or partition its output for shuffle.

Spark creates a task per data partition. An executor runs one or more tasks concurrently. The amount of parallelism is determined by the number of partitions. More partitions mean more tasks processing data in parallel.

Application Execution

This section briefly describes how data processing code is executed on a Spark cluster.

Terminology

Let's define a few terms first:

- *Shuffle.* A shuffle redistributes data among a cluster of nodes. It is an expensive operation because it involves moving data across a network. Note that a shuffle does not randomly redistribute data; it groups data elements into buckets based on some criteria. Each bucket forms a new partition.

- *Job.* A job is a set of computations that Spark performs to return results to a driver program. Essentially, it is an execution of a data processing algorithm on a Spark cluster. An application can launch multiple jobs. Exactly how a job is executed is covered later in this chapter.

- *Stage.* A stage is a collection of tasks. Spark splits a job into a DAG of stages. A stage may depend on another stage. For example, a job may be split into two stages, stage 0 and stage 1, where stage 1 cannot begin until stage 0 is completed. Spark groups tasks into stages using shuffle boundaries. Tasks that do not require a shuffle are grouped into the same stage. A task that requires its input data to be shuffled begins a new stage.

How an Application Works

With the definitions out of the way, I can now describe how a Spark application processes data in parallel across a cluster of nodes. When a Spark application is run, Spark connects to a cluster manager and acquires executors on the worker nodes. As mentioned earlier, a Spark application submits a data processing algorithm as a job. Spark splits a job into a directed acyclic graph (DAG) of stages. It then schedules the execution of these stages on the executors using a low-level scheduler provided by a cluster manager. The executors run the tasks submitted by Spark in parallel.

Every Spark application gets its own set of executors on the worker nodes. This design provides a few benefits. First, tasks from different applications are isolated from each other since they run in different JVM processes. A misbehaving task from one application cannot crash another Spark application. Second, scheduling of tasks becomes easier. Spark has to schedule the tasks belonging to only one application at a time. It does not have to handle the complexities of scheduling tasks from multiple concurrently running applications.

However, this design also has one disadvantage. Since applications run in separate JVM processes, they cannot easily share data. Even though they may be running on the same worker nodes, they cannot share data without writing it to disk. As previously mentioned, writing and reading data from disk are expensive operations. Therefore, applications sharing data through disk will experience performance issues.

Data Sources

Spark is essentially a computing framework for processing large datasets using a cluster of nodes. Unlike a database, it does not provide a storage system, but it works in conjunction with external storage systems. Generally, it is used with distributed storage systems that store large amounts of data.

Spark supports a variety of data sources. The data that you crunch through a Spark application can be in HDFS, HBase, Cassandra, Amazon S3, or any other Hadoop-supported data source. Any data source that works with Hadoop can be used with Spark core. One of the Spark libraries, Spark SQL, enables Spark to support even more data sources. Chapter 7 covers Spark SQL.

Compatibility with a Hadoop-supported data source is important. Organizations have made significant investments in Hadoop. Large amounts of data exist in HDFS and other Hadoop-supported storage systems. Spark does not require you to move or copy data from these sources to another storage system. Thus, migration from Hadoop MapReduce to Spark is not a forklift operation. This makes it easy to switch from Hadoop MapReduce to Spark. If you have an existing Hadoop cluster running MapReduce jobs, you can run Spark applications on the same cluster in parallel. You can convert existing MapReduce jobs to Spark jobs. Alternatively, if you are happy with the existing MapReduce applications and do not want to touch them, you can use Spark for new applications.

While Spark core has built-in support for Hadoop compatible storage systems, support for additional data sources can be easily added. For example, people have created Spark connectors for Cassandra, MongoDB, CouchDB, and other popular data sources.

Spark also supports local file systems. A Spark application can read input from and store output to a local file system. Although Spark is not needed if data can be read from a local file and processed on a single computer, this capability is useful for initial application development and debugging. It also makes it easy to learn Spark.

Application Programming Interface (API)

Spark makes its cluster computing capabilities available to an application in the form of a library. This library is written in Scala, but it provides an application programming interface (API) in multiple languages. At the time this book is being written, the Spark API is available in Scala, Java, Python, and R. You can develop a Spark application in any of these languages. Unofficial support for additional languages, such as Clojure, is also available.

The Spark API consists of two important abstractions: SparkContext and Resilient Distributed Datasets (RDDs). An application interacts with Spark using these two abstractions. These abstractions allow an application to connect to a Spark cluster and use the cluster resources. The next section discusses each abstraction and then looks at RDDs in more detail.

SparkContext

SparkContext is a class defined in the Spark library. It is the main entry point into the Spark library. It represents a connection to a Spark cluster. It is also required to create other important objects provided by the Spark API.

A Spark application must create an instance of the SparkContext class. Currently, an application can have only one active instance of SparkContext. To create another instance, it must first stop the active instance.

The SparkContext class provides multiple constructors. The simplest one does not take any arguments. An instance of the SparkContext class can be created as shown next.

```
val sc = new SparkContext()
```

In this case, SparkContext gets configuration settings such as the address of the Spark master, application name, and other settings from system properties. You can also provide configuration parameters to SparkContext programmatically using an instance of SparkConf, which is also a class defined in the Spark library. It can be used to set various Spark configuration parameters as shown next.

```
val config = new SparkConf().setMaster("spark://host:port").setAppName("big app")
val sc = new SparkContext(config)
```

In addition to providing explicit methods for configuring commonly used parameters such as the address of the Spark master, SparkConf provides a generic method for setting any parameter using a key-value pair. The input parameters that you can provide to SparkContext and SparkConf are covered in more detail in Chapter 4.

The variable sc created earlier is used in other examples in the rest of this chapter.

Resilient Distributed Datasets (RDD)

RDD represents a collection of partitioned data elements that can be operated on in parallel. It is the primary data abstraction mechanism in Spark. It is defined as an abstract class in the Spark library.

Conceptually, RDD is similar to a Scala collection, except that it represents a distributed dataset and it supports lazy operations. Lazy operations are discussed in detail later in this chapter.

The key characteristics of an RDD are briefly described in the following sections.

Immutable

An RDD is an immutable data structure. Once created, it cannot be modified in-place. Basically, an operation that modifies an RDD returns a new RDD.

Partitioned

Data represented by an RDD is split into partitions. These partitions are generally distributed across a cluster of nodes. However, when Spark is running on a single machine, all the partitions are on that machine.

Note that there is a mapping from RDD partitions to physical partitions of a dataset. RDD provides an abstraction for data stored in distributed data sources, which generally partition data and distribute it across a cluster of nodes. For example, HDFS stores data in partitions or blocks, which are distributed across a cluster of computers. By default, there is one-to-one mapping between an RDD partition and a HDFS file partition. Other distributed data sources, such as Cassandra, also partition and distribute data across a cluster of nodes. However, multiple Cassandra partitions are mapped to a single RDD partition.

Fault Tolerant

RDD is designed to be fault tolerant. An RDD represents data distributed across a cluster of nodes and a node can fail. As previously discussed, the probability of a node failing is proportional to the number of nodes in a cluster. The larger a cluster, the higher the probability that some node will fail on any given day.

RDD automatically handles node failures. When a node fails, and partitions stored on that node become inaccessible, Spark reconstructs the lost RDD partitions on another node. Spark stores lineage information for each RDD. Using this lineage information, it can recover parts of an RDD or even an entire RDD in the event of node failures.

Interface

It is important to remember that RDD is an interface for processing data. It is defined as an abstract class in the Spark library. RDD provides a uniform interface for processing data from a variety of data sources, such as HDFS, HBase, Cassandra, and others. The same interface can also be used to process data stored in memory across a cluster of nodes.

Spark provides concrete implementation classes for representing different data sources. Examples of concrete RDD implementation classes include `HadoopRDD`, `ParallelCollectionRDD`, `JdbcRDD`, and `CassandraRDD`. They all support the base RDD interface.

Strongly Typed

The RDD class definition has a type parameter. This allows an RDD to represent data of different types. It is a distributed collection of homogenous elements, which can be of type `Integer`, `Long`, `Float`, `String`, or a custom type defined by an application developer. Thus, an application always works with an RDD of some type. It can be an RDD of `Integer`, `Long`, `Float`, `Double`, `String`, or a custom type.

In Memory

Spark's in-memory cluster computing capabilities was covered earlier in this chapter. The RDD class provides the API for enabling in-memory cluster computing. Spark allows RDDs to be cached or persisted in memory. As mentioned, operations on an RDD cached in memory are orders of magnitude faster than those operating on a non-cached RDD.

Creating an RDD

Since RDD is an abstract class, you cannot create an instance of the RDD class directly. The `SparkContext` class provides factory methods to create instances of concrete implementation classes. An RDD can also be created from another RDD by applying a transformation to it. As discussed earlier, RDDs are immutable. Any operation that modifies an RDD returns a new RDD with the modified data.

The methods commonly used to create an RDD are briefly described in this section. In the following code examples, `sc` is an instance of the `SparkContext` class. You learned how to create it earlier in this chapter.

parallelize

This method creates an RDD from a local Scala collection. It partitions and distributes the elements of a Scala collection and returns an RDD representing those elements. This method is generally not used in a production application, but useful for learning Spark.

```
val xs = (1 to 10000).toList
val rdd = sc.parallelize(xs)
```

textFile

The textFile method creates an RDD from a text file. It can read a file or multiple files in a directory stored on a local file system, HDFS, Amazon S3, or any other Hadoop-supported storage system. It returns an RDD of Strings, where each element represents a line in the input file.

```
val rdd = sc.textFile("hdfs://namenode:9000/path/to/file-or-directory")
```

The preceding code will create an RDD from a file or directory stored on HDFS.

The textFile method can also read compressed files. In addition, it supports wildcards as an argument for reading multiple files from a directory. An example is shown next.

```
val rdd = sc.textFile("hdfs://namenode:9000/path/to/directory/*.gz")
```

The textFile method takes an optional second argument, which can be used to specify the number of partitions. By default, Spark creates one RDD partition for each file block. You can specify a higher number of partitions for increasing parallelism; however, a fewer number of partitions than file blocks is not allowed.

wholeTextFiles

This method reads all text files in a directory and returns an RDD of key-value pairs. Each key-value pair in the returned RDD corresponds to a single file. The key part stores the path of a file and the value part stores the content of a file. This method can also read files stored on a local file system, HDFS, Amazon S3, or any other Hadoop-supported storage system.

```
val rdd = sc.wholeTextFiles("path/to/my-data/*.txt")
```

sequenceFile

The sequenceFile method reads key-value pairs from a sequence file stored on a local file system, HDFS, or any other Hadoop-supported storage system. It returns an RDD of key-value pairs. In addition to providing the name of an input file, you have to specify the data types for the keys and values as type parameters when you call this method.

```
val rdd = sc.sequenceFile[String, String]("some-file")
```

RDD Operations

Spark applications process data using the methods defined in the RDD class or classes derived from it. These methods are also referred to as operations. Since Scala allows a method to be used with operator notation, the RDD methods are also sometimes referred to as *operators*.

The beauty of Spark is that the same RDD methods can be used to process data ranging in size from a few bytes to several petabytes. In addition, a Spark application can use the same methods to process datasets stored on either a distributed storage system or a local file system. This flexibility allows a developer to develop, debug and test a Spark application on a single machine and deploy it on a large cluster without making any code change.

RDD operations can be categorized into two types: transformation and action. A transformation creates a new RDD. An action returns a value to a driver program.

Transformations

A transformation method of an RDD creates a new RDD by performing a computation on the source RDD. This section discusses the commonly used RDD transformations.

RDD transformations are conceptually similar to Scala collection methods. The key difference is that the Scala collection methods operate on data that can fit in the memory of a single machine, whereas RDD methods can operate on data distributed across a cluster of nodes. Another important difference is that RDD transformations are lazy, whereas Scala collection methods are strict. This topic is discussed in more detail later in this chapter.

map

The map method is a higher-order method that takes a function as input and applies it to each element in the source RDD to create a new RDD. The input function to map must take a single input parameter and return a value.

```
val lines = sc.textFile("...")
val lengths = lines map { l => l.length}
```

filter

The filter method is a higher-order method that takes a Boolean function as input and applies it to each element in the source RDD to create a new RDD. A Boolean function takes an input and returns true or false. The filter method returns a new RDD formed by selecting only those elements for which the input Boolean function returned true. Thus, the new RDD contains a subset of the elements in the original RDD.

```
val lines = sc.textFile("...")
val longLines = lines filter { l => l.length > 80}
```

flatMap

The flatMap method is a higher-order method that takes an input function, which returns a sequence for each input element passed to it. The flatMap method returns a new RDD formed by flattening this collection of sequence.

```
val lines = sc.textFile("...")
val words = lines flatMap { l => l.split(" ")}
```

mapPartitions

The higher-order `mapPartitions` method allows you to process data at a partition level. Instead of passing one element at a time to its input function, `mapPartitions` passes a partition in the form of an iterator. The input function to the `mapPartitions` method takes an iterator as input and returns another iterator as output. The `mapPartitions` method returns new RDD formed by applying a user-specified function to each partition of the source RDD.

```
val lines = sc.textFile("...")
val lengths = lines mapPartitions { iter => iter.map { l => l.length}}
```

union

The `union` method takes an RDD as input and returns a new RDD that contains the union of the elements in the source RDD and the RDD passed to it as an input.

```
val linesFile1 = sc.textFile("...")
val linesFile2 = sc.textFile("...")
val linesFromBothFiles = linesFile1.union(linesFile2)
```

intersection

The `intersection` method takes an RDD as input and returns a new RDD that contains the intersection of the elements in the source RDD and the RDD passed to it as an input.

```
val linesFile1 = sc.textFile("...")
val linesFile2 = sc.textFile("...")
val linesPresentInBothFiles = linesFile1.intersection(linesFile2)
```

Here is another example.

```
val mammals = sc.parallelize(List("Lion", "Dolphin", "Whale"))
val aquatics =sc.parallelize(List("Shark", "Dolphin", "Whale"))
val aquaticMammals = mammals.intersection(aquatics)
```

subtract

The `subtract` method takes an RDD as input and returns a new RDD that contains elements in the source RDD but not in the input RDD.

```
val linesFile1 = sc.textFile("...")
val linesFile2 = sc.textFile("...")
val linesInFile1Only = linesFile1.subtract(linesFile2)
```

Here is another example.

```
val mammals = sc.parallelize(List("Lion", "Dolphin", "Whale"))
val aquatics =sc.parallelize(List("Shark", "Dolphin", "Whale"))
val fishes = aquatics.subtract(mammals)
```

distinct

The distinct method of an RDD returns a new RDD containing the distinct elements in the source RDD.

```
val numbers = sc.parallelize(List(1, 2, 3, 4, 3, 2, 1))
val uniqueNumbers = numbers.distinct
```

cartesian

The cartesian method of an RDD takes an RDD as input and returns an RDD containing the cartesian product of all the elements in both RDDs. It returns an RDD of ordered pairs, in which the first element comes from the source RDD and the second element is from the input RDD. The number of elements in the returned RDD is equal to the product of the source and input RDD lengths.

This method is similar to a cross join operation in SQL.

```
val numbers = sc.parallelize(List(1, 2, 3, 4))
val alphabets = sc.parallelize(List("a", "b", "c", "d"))
val cartesianProduct = numbers.cartesian(alphabets)
```

zip

The zip method takes an RDD as input and returns an RDD of pairs, where the first element in a pair is from the source RDD and second element is from the input RDD. Unlike the cartesian method, the RDD returned by zip has the same number of elements as the source RDD. Both the source RDD and the input RDD must have the same length. In addition, both RDDs are assumed to have same number of partitions and same number of elements in each partition.

```
val numbers = sc.parallelize(List(1, 2, 3, 4))
val alphabets = sc.parallelize(List("a", "b", "c", "d"))
val zippedPairs = numbers.zip(alphabets)
```

zipWithIndex

The zipWithIndex method zips the elements of the source RDD with their indices and returns an RDD of pairs.

```
val alphabets = sc.parallelize(List("a", "b", "c", "d"))
val alphabetsWithIndex = alphabets.zip
```

groupBy

The higher-order groupBy method groups the elements of an RDD according to a user specified criteria. It takes as input a function that generates a key for each element in the source RDD. It applies this function to all the elements in the source RDD and returns an RDD of pairs. In each returned pair, the first item is a key and the second item is a collection of the elements mapped to that key by the input function to the groupBy method.

Note that the groupBy method is an expensive operation since it may shuffle data.

Consider a CSV file that stores the name, age, gender, and zip code of customers of a company. The following code groups customers by their zip codes.

```scala
case class Customer(name: String, age: Int, gender: String, zip: String)
val lines = sc.textFile("...")
val customers = lines map { l => {
                val a = l.split(",")
                Customer(a(0), a(1).toInt, a(2), a(3))
             }
          }
val groupByZip = customers.groupBy { c => c.zip}
```

keyBy

The keyBy method is similar to the groupBy method. It a higher-order method that takes as input a function that returns a key for any given element in the source RDD. The keyBy method applies this function to all the elements in the source RDD and returns an RDD of pairs. In each returned pair, the first item is a key and the second item is an element that was mapped to that key by the input function to the keyBy method. The RDD returned by keyBy will have the same number of elements as the source RDD.

The difference between groupBy and keyBy is that the second item in a returned pair is a collection of elements in the first case, while it is a single element in the second case.

```scala
case class Person(name: String, age: Int, gender: String, zip: String)
val lines = sc.textFile("...")
val people = lines map { l => {
                val a = l.split(",")
                Person(a(0), a(1).toInt, a(2), a(3))
             }
          }
val keyedByZip = people.keyBy { p => p.zip}
```

sortBy

The higher-order sortBy method returns an RDD with sorted elements from the source RDD. It takes two input parameters. The first input is a function that generates a key for each element in the source RDD. The second argument allows you to specify ascending or descending order for sort.

```scala
val numbers = sc.parallelize(List(3,2, 4, 1, 5))
val sorted = numbers.sortBy(x => x, true)
```

Here is another example.

```scala
case class Person(name: String, age: Int, gender: String, zip: String)
val lines = sc.textFile("...")
val people = lines map { l => {
                val a = l.split(",")
                Person(a(0), a(1).toInt, a(2), a(3))
             }
          }
val sortedByAge = people.sortBy( p => p.age, true)
```

pipe

The pipe method allows you to execute an external program in a forked process. It captures the output of the external program as a String and returns an RDD of Strings.

randomSplit

The randomSplit method splits the source RDD into an array of RDDs. It takes the weights of the splits as input.

```
val numbers = sc.parallelize((1 to 100).toList)
val splits = numbers.randomSplit(Array(0.6, 0.2, 0.2))
```

coalesce

The coalesce method reduces the number of partitions in an RDD. It takes an integer input and returns a new RDD with the specified number of partitions.

```
val numbers = sc.parallelize((1 to 100).toList)
val numbersWithOnePartition = numbers.coalesce(1)
```

The coalesce method should be used with caution since reducing the number of partitions reduces the parallelism of a Spark application. It is generally useful for consolidating partitions with few elements. For example, an RDD may have too many sparse partitions after a filter operation. Reducing the partitions may provide performance benefit in such a case.

repartition

The repartition method takes an integer as input and returns an RDD with specified number of partitions. It is useful for increasing parallelism. It redistributes data, so it is an expensive operation.

The coalesce and repartition methods look similar, but the first one is used for reducing the number of partitions in an RDD, while the second one is used to increase the number of partitions in an RDD.

```
val numbers = sc.parallelize((1 to 100).toList)
val numbersWithOnePartition = numbers.repartition(4)
```

sample

The sample method returns a sampled subset of the source RDD. It takes three input parameters. The first parameter specifies the replacement strategy. The second parameter specifies the ratio of the sample size to source RDD size. The third parameter, which is optional, specifies a random seed for sampling.

```
val numbers = sc.parallelize((1 to 100).toList)
val sampleNumbers = numbers.sample(true, 0.2)
```

Transformations on RDD of key-value Pairs

In addition to the transformations described in the previous sections, RDDs of key-value pairs support a few other transformations. The commonly used transformations available for only RDDs of key-value pairs are briefly described next.

keys

The keys method returns an RDD of only the keys in the source RDD.

```
val kvRdd = sc.parallelize(List(("a", 1), ("b", 2), ("c", 3)))
val keysRdd = kvRdd.keys
```

values

The values method returns an RDD of only the values in the source RDD.

```
val kvRdd = sc.parallelize(List(("a", 1), ("b", 2), ("c", 3)))
val valuesRdd = kvRdd.values
```

mapValues

The mapValues method is a higher-order method that takes a function as input and applies it to each value in the source RDD. It returns an RDD of key-value pairs. It is similar to the map method, except that it applies the input function only to each value in the source RDD, so the keys are not changed. The returned RDD has the same keys as the source RDD.

```
val kvRdd = sc.parallelize(List(("a", 1), ("b", 2), ("c", 3)))
val valuesDoubled = kvRdd mapValues { x => 2*x}
```

join

The join method takes an RDD of key-value pairs as input and performs an inner join on the source and input RDDs. It returns an RDD of pairs, where the first element in a pair is a key found in both source and input RDD and the second element is a tuple containing values mapped to that key in the source and input RDD.

```
val pairRdd1 = sc.parallelize(List(("a", 1), ("b",2), ("c",3)))
val pairRdd2 = sc.parallelize(List(("b", "second"), ("c","third"), ("d","fourth")))
val joinRdd = pairRdd1.join(pairRdd2)
```

leftOuterJoin

The leftOuterJoin method takes an RDD of key-value pairs as input and performs a left outer join on the source and input RDD. It returns an RDD of key-value pairs, where the first element in a pair is a key from source RDD and the second element is a tuple containing value from source RDD and optional value from the input RDD. An optional value from the input RDD is represented with Option type.

```
val pairRdd1 = sc.parallelize(List(("a", 1), ("b",2), ("c",3)))
val pairRdd2 = sc.parallelize(List(("b", "second"), ("c","third"), ("d","fourth")))
val leftOuterJoinRdd = pairRdd1.leftOuterJoin(pairRdd2)
```

rightOuterJoin

The rightOuterJoin method takes an RDD of key-value pairs as input and performs a right outer join on the source and input RDD. It returns an RDD of key-value pairs, where the first element in a pair is a key from input RDD and the second element is a tuple containing optional value from source RDD and value from input RDD. An optional value from the source RDD is represented with the Option type.

```scala
val pairRdd1 = sc.parallelize(List(("a", 1), ("b",2), ("c",3)))
val pairRdd2 = sc.parallelize(List(("b", "second"), ("c","third"), ("d","fourth")))
val rightOuterJoinRdd = pairRdd1.rightOuterJoin(pairRdd2)
```

fullOuterJoin

The fullOuterJoin method takes an RDD of key-value pairs as input and performs a full outer join on the source and input RDD. It returns an RDD of key-value pairs.

```scala
val pairRdd1 = sc.parallelize(List(("a", 1), ("b",2), ("c",3)))
val pairRdd2 = sc.parallelize(List(("b", "second"), ("c","third"), ("d","fourth")))
val fullOuterJoinRdd = pairRdd1.fullOuterJoin(pairRdd2)
```

sampleByKey

The sampleByKey method returns a subset of the source RDD sampled by key. It takes the sampling rate for each key as input and returns a sample of the source RDD.

```scala
val pairRdd = sc.parallelize(List(("a", 1), ("b",2), ("a", 11),("b",22),("a", 111), ("b",222)))
val sampleRdd = pairRdd.sampleByKey(true, Map("a"-> 0.1, "b"->0.2))
```

subtractByKey

The subtractByKey method takes an RDD of key-value pairs as input and returns an RDD of key-value pairs containing only those keys that exist in the source RDD, but not in the input RDD.

```scala
val pairRdd1 = sc.parallelize(List(("a", 1), ("b",2), ("c",3)))
val pairRdd2 = sc.parallelize(List(("b", "second"), ("c","third"), ("d","fourth")))
val resultRdd = pairRdd1.subtractByKey(pairRdd2)
```

groupByKey

The groupByKey method returns an RDD of pairs, where the first element in a pair is a key from the source RDD and the second element is a collection of all the values that have the same key. It is similar to the groupBy method that we saw earlier. The difference is that groupBy is a higher-order method that takes as input a function that returns a key for each element in the source RDD. The groupByKey method operates on an RDD of key-value pairs, so a key generator function is not required as input.

```scala
val pairRdd = sc.parallelize(List(("a", 1), ("b",2), ("c",3), ("a", 11), ("b",22), ("a",111)))
val groupedRdd = pairRdd.groupByKey()
```

The groupByKey method should be avoided. It is an expensive operation since it may shuffle data. For most use cases, better alternatives are available.

reduceByKey

The higher-order reduceByKey method takes an associative binary operator as input and reduces values with the same key to a single value using the specified binary operator.

A binary operator takes two values as input and returns a single value as output. An associative operator returns the same result regardless of the grouping of the operands.

The reduceByKey method can be used for aggregating values by key. For example, it can be used for calculating sum, product, minimum or maximum of all the values mapped to the same key.

```
val pairRdd = sc.parallelize(List(("a", 1), ("b",2), ("c",3), ("a", 11), ("b",22), ("a",111)))
val sumByKeyRdd = pairRdd.reduceByKey((x,y) => x+y)
val minByKeyRdd = pairRdd.reduceByKey((x,y) => if (x < y) x else y)
```

The reduceByKey method is a better alternative than groupByKey for key-based aggregations or merging.

Actions

Actions are RDD methods that return a value to a driver program. This section discusses the commonly used RDD actions.

collect

The collect method returns the elements in the source RDD as an array. This method should be used with caution since it moves data from all the worker nodes to the driver program. It can crash the driver program if called on a very large RDD.

```
val rdd = sc.parallelize((1 to 10000).toList)
val filteredRdd = rdd filter { x => (x % 1000) == 0 }
val filterResult = filteredRdd.collect
```

count

The count method returns a count of the elements in the source RDD.

```
val rdd = sc.parallelize((1 to 10000).toList)
val total = rdd.count
```

countByValue

The countByValue method returns a count of each unique element in the source RDD. It returns an instance of the Map class containing each unique element and its count as a key-value pair.

```
val rdd = sc.parallelize(List(1, 2, 3, 4, 1, 2, 3, 1, 2, 1))
val counts = rdd.countByValue
```

first

The first method returns the first element in the source RDD.

```
val rdd = sc.parallelize(List(10, 5, 3, 1))
val firstElement = rdd.first
```

max

The max method returns the largest element in an RDD.

```
val rdd = sc.parallelize(List(2, 5, 3, 1))
val maxElement = rdd.max
```

min

The min method returns the smallest element in an RDD.

```
val rdd = sc.parallelize(List(2, 5, 3, 1))
val minElement = rdd.min
```

take

The take method takes an integer N as input and returns an array containing the first N element in the source RDD.

```
val rdd = sc.parallelize(List(2, 5, 3, 1, 50, 100))
val first3 = rdd.take(3)
```

takeOrdered

The takeOrdered method takes an integer N as input and returns an array containing the N smallest elements in the source RDD.

```
val rdd = sc.parallelize(List(2, 5, 3, 1, 50, 100))
val smallest3 = rdd.takeOrdered(3)
```

top

The top method takes an integer N as input and returns an array containing the N largest elements in the source RDD.

```
val rdd = sc.parallelize(List(2, 5, 3, 1, 50, 100))
val largest3 = rdd.top(3)
```

fold

The higher-order fold method aggregates the elements in the source RDD using the specified neutral zero value and an associative binary operator. It first aggregates the elements in each RDD partition and then aggregates the results from each partition.

The neutral zero value depends on the RDD type and the aggregation operation. For example, if you want to sum all the elements in an RDD of Integers, the neutral zero value should be 0. Instead, if you want to calculate the products of all the elements in an RDD of Integers, the neutral zero value should be 1.

```
val numbersRdd = sc.parallelize(List(2, 5, 3, 1))
val sum = numbersRdd.fold(0) ((partialSum, x) => partialSum + x)
val product = numbersRdd.fold(1) ((partialProduct, x) => partialProduct * x)
```

reduce

The higher-order reduce method aggregates the elements of the source RDD using an associative and commutative binary operator provided to it. It is similar to the fold method; however, it does not require a neutral zero value.

```
val numbersRdd = sc.parallelize(List(2, 5, 3, 1))
val sum = numbersRdd.reduce ((x, y) => x + y)
val product = numbersRdd.reduce((x, y) => x * y)
```

Actions on RDD of key-value Pairs

RDDs of key-value pairs support a few additional actions, which are briefly described next.

countByKey

The countByKey method counts the occurrences of each unique key in the source RDD. It returns a Map of key-count pairs.

```
val pairRdd = sc.parallelize(List(("a", 1), ("b", 2), ("c", 3), ("a", 11), ("b", 22), ("a", 1)))
val countOfEachKey = pairRdd.countByKey
```

lookup

The lookup method takes a key as input and returns a sequence of all the values mapped to that key in the source RDD.

```
val pairRdd = sc.parallelize(List(("a", 1), ("b", 2), ("c", 3), ("a", 11), ("b", 22), ("a", 1)))
val values = pairRdd.lookup("a")
```

Actions on RDD of Numeric Types

RDDs containing data elements of type Integer, Long, Float, or Double support a few additional actions that are useful for statistical analysis. The commonly used actions from this group are briefly described next.

mean

The mean method returns the average of the elements in the source RDD.

```
val numbersRdd = sc.parallelize(List(2, 5, 3, 1))
val mean = numbersRdd.mean
```

stdev

The stdev method returns the standard deviation of the elements in the source RDD.

```
val numbersRdd = sc.parallelize(List(2, 5, 3, 1))
val stdev = numbersRdd.stdev
```

sum

The sum method returns the sum of the elements in the source RDD.

```
val numbersRdd = sc.parallelize(List(2, 5, 3, 1))
val sum = numbersRdd.sum
```

variance

The variance method returns the variance of the elements in the source RDD.

```
val numbersRdd = sc.parallelize(List(2, 5, 3, 1))
val variance = numbersRdd.variance
```

Saving an RDD

Generally, after data is processed, results are saved on disk. Spark allows an application developer to save an RDD to any Hadoop-supported storage system. An RDD saved to disk can be used by another Spark or MapReduce application.

This section presents commonly used RDD methods to save an RDD to a file.

saveAsTextFile

The saveAsTextFile method saves the elements of the source RDD in the specified directory on any Hadoop-supported file system. Each RDD element is converted to its string representation and stored as a line of text.

```
val numbersRdd = sc.parallelize((1 to 10000).toList)
val filteredRdd = numbersRdd filter { x => x % 1000 == 0}
filteredRdd.saveAsTextFile("numbers-as-text")
```

saveAsObjectFile

The `saveAsObjectFile` method saves the elements of the source RDD as serialized Java objects in the specified directory.

```
val numbersRdd = sc.parallelize((1 to 10000).toList)
val filteredRdd = numbersRdd filter { x => x % 1000 == 0}
filteredRdd.saveAsObjectFile("numbers-as-object")
```

saveAsSequenceFile

The `saveAsSequenceFile` method saves an RDD of key-value pairs in SequenceFile format. An RDD of key-value pairs can also be saved in text format using the `saveAsTextFile`.

```
val pairs = (1 to 10000).toList map {x => (x, x*2)}
val pairsRdd = sc.parallelize(pairs)
val filteredPairsRdd = pairsRdd filter { case (x, y) => x % 1000 ==0 }
filteredPairsRdd.saveAsSequenceFile("pairs-as-sequence")
filteredPairsRdd.saveAsTextFile("pairs-as-text")
```

Note that all of the preceding methods take a directory name as an input parameter and create one file for each RDD partition in the specified directory. This design is both efficient and fault tolerant. Since each partition is stored in a separate file, Spark launches multiple tasks and runs them in parallel to write an RDD to a file system. It also helps makes the file writing process fault tolerant. If a task writing a partition to a file fails, Spark creates another task, which rewrites the file that was created by the failed task.

Lazy Operations

RDD creation and transformation methods are lazy operations. Spark does not immediately perform any computation when an application calls a method that return an RDD. For example, when you read a file from HDFS using `textFile` method of `SparkContext`, Spark does not immediately read the file from disk. Similarly, RDD transformations, which return a new RDD, are lazily computed. Spark just keeps track of transformations applied to an RDD.

Let's consider the following sample code.

```
val lines = sc.textFile("...")
val errorLines = lines filter { l => l.contains("ERROR")}
val warningLines = lines filter { l => l.contains("WARN")}
```

These three lines of code will seem to execute very quickly, even if you pass a file containing 100 terabytes of data to the `textFile` method. The reason is that the `textFile` method does not actually read a file right when you call it. Similarly, the `filter` method does not immediately iterate through all the elements in the source RDD.

Spark just makes a note of how an RDD was created and the transformations applied to it to create child RDDs. Thus, it maintains lineage information for each RDD. It uses this lineage information to construct or reconstruct an RDD when required.

If RDD creation and transformations are lazy operations, when does Spark actually read data and compute transformations? The next section answers this question.

Action Triggers Computation

RDD transformations are computed when an application calls an action method of an RDD or saves an RDD to a storage system. Saving an RDD to a storage system is considered as an action, even though it does not return a value to the driver program.

When an application calls an RDD action method or saves an RDD, it triggers a chain reaction in Spark. At that point, Spark attempts to create the RDD whose action method was called. If that RDD was generated from a file, Spark reads that file into the memory of the worker nodes. If it is a child RDD created by a transformation of another RDD, then Spark attempts to first create the parent RDD. This process continues until Spark finds the root RDD. It then performs all the transformations required to generate the RDD whose action method was called. Finally, it performs the computations to generate the result that the action method returns to the driver program.

Lazy transformations enable Spark to run RDD computations efficiently. By delaying computations until an application needs the result of an action, Spark can optimize RDD operations. It pipelines operations and avoids unnecessary transfer of data over the network.

Caching

Besides storing data in memory, caching an RDD play another important role. As mentioned earlier, an RDD is created by either reading data from a storage system or by applying a transformation to an existing RDD. By default, when an action method of an RDD is called, Spark creates that RDD from its parents, which may require creation of the parent RDDs, and so on. This process continues until Spark gets to the root RDD, which Spark creates by reading data from a storage system. This happens every time an action method is called. Thus, by default, every time an action method is called, Spark traverses the lineage tree of an RDD and computes all the transformations to obtain the RDD whose action method was called.

Consider the following example.

```
val logs = sc.textFile("path/to/log-files")
val errorLogs = logs filter { l => l.contains("ERROR")}
val warningLogs = logs filter { l => l.contains("WARN")}
val errorCount = errorLogs.count
val warningCount =  warningLogs.count
```

Even though the preceding code calls the textFile method only once, the log files will be read twice from disk since there are two calls to the action method count. The log files will read first when errorLogs.count is called. It will be again read when warningLogs.count is called. It is a simple example. A real-world application may have a lot more calls to different transformations and actions.

When an RDD is cached, Spark computes all the transformations up to that point and creates a checkpoint for that RDD. To be more accurate, it happens the first time an action is called on a cached RDD. Similar to transformation methods, caching is lazy.

When an application caches an RDD, Spark does not immediately compute the RDD and store it in memory. It materializes an RDD in memory the first time an action is called on the cached RDD. Thus, the first action called after an RDD is cached does not benefit from caching. Only subsequent actions benefit from caching. Since subsequent calls to an action method no longer have to start from reading data from a storage system, they generally execute much faster. Thus, an application that does just one pass over data will not benefit from caching. Only applications that iterate over the same data more than once will benefit from RDD caching.

When an application caches an RDD in memory, Spark stores it in the executor memory on each worker node. Each executor stores in memory the RDD partitions that it computes.

RDD Caching Methods

The RDD class provides two methods to cache an RDD: cache and persist.

cache

The cache method stores an RDD in the memory of the executors across a cluster. It essentially materializes an RDD in memory.

The example shown earlier can be optimized using the cache method as shown next.

```
val logs = sc.textFile("path/to/log-files")
val errorsAndWarnings = logs filter { l => l.contains("ERROR") || l.contains("WARN")}
errorsAndWarnings.cache()
val errorLogs = errorsAndWarnings filter { l => l.contains("ERROR")}
val warningLogs = errorsAndWarnings filter { l => l.contains("WARN")}
val errorCount = errorLogs.count
val warningCount =  warningLogs.count
```

persist

The persist method is a generic version of the cache method. It allows an RDD to be stored in memory, disk, or both. It optionally takes a storage level as an input parameter. If persist is called without any parameter, its behavior is identical to that of the cache method.

```
val lines = sc.textFile("...")
lines.persist()
```

The persist method supports the following common storage options:

- MEMORY_ONLY: When an application calls the persist method with the MEMORY_ONLY flag, Spark stores RDD partitions in memory on the worker nodes using deserialized Java objects. If an RDD partition does not fit in memory on a worker node, it is computed on the fly when needed.

  ```
  val lines = sc.textFile("...")
  lines.persist(MEMORY_ONLY)
  ```

- DISK_ONLY: If persist is called with the DISK_ONLY flag, Spark materializes RDD partitions and stores them in a local file system on each worker node. This option can be used to persist intermediate RDDs so that subsequent actions do not have to start computation from the root RDD.

- MEMORY_AND_DISK: In this case, Spark stores as many RDD partitions in memory as possible and stores the remaining partitions on disk.

- MEMORY_ONLY_SER: In this case, Spark stores RDD partitions in memory as serialized Java objects. A serialized Java object consumes less memory, but is more CPU-intensive to read. This option allows a trade-off between memory consumption and CPU utilization.

- MEMORY_AND_DISK_SER: Spark stores in memory as serialized Java objects as many RDD partitions as possible. The remaining partitions are saved on disk.

RDD Caching Is Fault Tolerant

Fault tolerance is important in a distributed environment. Earlier you learned how Spark automatically moves a compute job to another node when a node fails. Spark's RDD caching mechanism is also fault tolerant.

A Spark application will not crash if a node with cached RDD partitions fails. Spark automatically recreates and caches the partitions stored on the failed node on another node. Spark uses RDD lineage information to recompute lost cached partitions.

Cache Memory Management

Spark automatically manages cache memory using LRU (least recently used) algorithm. It removes old RDD partitions from cache memory when needed. In addition, the RDD API includes a method called unpersist(). An application can call this method to manually remove RDD partitions from memory.

Spark Jobs

RDD operations, including transformation, action and caching methods form the basis of a Spark application. Essentially, RDDs describe the Spark programming model. Now that we have covered the programming model, we will discuss how it all comes together in a Spark application.

A job is a set of computations that Spark performs to return the results of an action to a driver program. An application can launch one or more jobs. It launches a job by calling an action method of an RDD. Thus, an action method triggers a job. If an action is called for an RDD that is not cached or a descendant of a cached RDD, a job starts with the reading of data from a storage system. However, if an action is called for an RDD that is cached or a descendent of a cached RDD, a job begins from the point at which the RDD or its ancestor RDD was cached. Next, Spark applies the transformations required to create the RDD whose action method was called. Finally, it performs the computations specified by the action. A job is completed when a result is returned to a driver program.

When an application calls an RDD action method, Spark creates a DAG of task stages. It groups tasks into stages using shuffle boundaries. Tasks that do not require a shuffle are grouped into the same stage. A task that requires its input data to be shuffled begins a new stage.

A stage can have one or more tasks. Spark submits tasks to the executors, which run the tasks in parallel. Tasks are scheduled on nodes based on data locality. If a node fails while working on a task, Spark resubmits task to another node.

Shared Variables

Spark uses a shared-nothing architecture. Data is partitioned across a cluster of nodes and each node in a cluster has its own CPU, memory, and storage resources. There is no global memory space that can be shared by the tasks. The driver program and job tasks share data through messages.

For example, if a function argument to an RDD operator references a variable in the driver program, Spark sends a copy of that variable along with a task to the executors. Each task gets its own copy of the variable and uses it as a read-only variable. Any update made to that variable by a task remains local. Changes are not propagated back to the driver program. In addition, Spark ships that variable to a worker node at the beginning of every stage.

This default behavior can be inefficient for some applications. In one use case, the driver program shares a large lookup table with the tasks in a job and the job involves several stages. By default, Spark automatically sends the driver variables referenced by a task to each executor; however, it does this for each stage. Thus, if the lookup table holds 100 MB data and the job involves ten stages, Spark will send the same 100 MB data to each worker node ten times.

Another use case involves the ability to update a global variable in each task running on different nodes. By default, updates made to a variable by a task are not propagated back to the driver program.

Spark supports the concept of shared variables for these use cases.

Broadcast Variables

Broadcast variables enable a Spark application to optimize sharing of data between the driver program and the tasks executing a job. Spark sends a broadcast variable to a worker node only once and caches it in deserialized form as a read-only variable in executor memory. In addition, it uses a more efficient algorithm to distribute broadcast variables.

Note that a broadcast variable is useful if a job consists of multiple stages and tasks across stages reference the same driver variable. It is also useful if you do not want the performance hit from having to deserialize a variable before running each task. By default, Spark caches a shipped variable in the executor memory in serialized form and deserializes it before running each task.

The SparkContext class provides a method named broadcast for creating a broadcast variable. It takes the variable to be broadcasted as an argument and returns an instance of the Broadcast class. A task must use the value method of a Broadcast object to access a broadcasted variable.

Consider an application where we want to generate transaction details from e-commerce transactions. In a real-world application, there would be a master customer table, a master item table, and transactions table. To keep the example simple, the input data is instead created in the code itself using simple data structures.

```
case class Transaction(id: Long, custId: Int, itemId: Int)
case class TransactionDetail(id: Long, custName: String, itemName: String)

val customerMap = Map(1 -> "Tom", 2 -> "Harry")
val itemMap = Map(1 -> "Razor", 2 -> "Blade")

val transactions = sc.parallelize(List(Transaction(1, 1, 1), Transaction(2, 1, 2)))

val bcCustomerMap = sc.broadcast(customerMap)
val bcItemMap = sc.broadcast(itemMap)

val transactionDetails = transactions.map{t => TransactionDetail(
                    t.id, bcCustomerMap.value(t.custId), bcItemMap.value(t.itemId))}
transactionDetails.collect
```

The use of broadcast variables enabled us to implement an efficient join between the customer, item and transaction dataset. We could have used the join operator from the RDD API, but that would shuffle customer, item, and transaction data over the network. Using broadcast variables, we instructed Spark to send customer and item data to each node only once and replaced an expensive join operation with a simple map operation.

Accumulators

An accumulator is an add-only variable that can be updated by tasks running on different nodes and read by the driver program. It can be used to implement counters and aggregations. Spark comes pre-packaged with accumulators of numeric types and it supports creation of custom accumulators.

The SparkContext class provides a method named accumulator for creating an accumulator variable. It takes two arguments. The first argument is the initial value for the accumulator and the second argument, which is optional, is a name for displaying in the Spark UI. It returns an instance of the Accumulator class, which provides the operators for working with an accumulator variable. Tasks can only add a value to an accumulator variable using the add method or += operator. Only the driver program can read an accumulator's value using it value method.

Let's consider an application that needs to filter out and count the number of invalid customer identifiers in a customer table. In a real-world application, we will read the input data from disk and write the filtered data back to another file disk. To keep the example simple, we will skip reading and writing to disk.

```
case class Customer(id: Long, name: String)
val customers = sc.parallelize(List(
                    Customer(1, "Tom"),
                    Customer(2, "Harry"),
                    Customer(-1, "Paul")))
val badIds = sc.accumulator(0, "Bad id accumulator")
val validCustomers = customers.filter(c => if (c.id < 0) {
                                           badIds += 1
                                           false
                                     } else true
                              )
val validCustomerIds = validCustomers.count
val invalidCustomerIds = badIds.value
```

Accumulators should be used with caution. Updates to an accumulator within a transformation are not guaranteed to be performed exactly once. If a task or stage is re-executed, each task's update will be applied more than once.

In addition, the update statements are not executed until an RDD action method is called. RDD transformations are lazy; accumulator updates within a transformation are not executed right away. Therefore, if a driver program uses the value of an accumulator before an action is called, it will get the wrong value.

Summary

Spark is a fast, scalable, fault-tolerant, general-purpose, in-memory cluster computing framework. A Spark application can be up to 100 times faster than an application built with Hadoop MapReduce.

Spark is not only faster than MapReduce but also easier to use than MapReduce. With an expressive API in Java, Python Scala, and R, Spark makes it easier to develop distributed big data applications. A developer can be five to ten times more productive with Spark.

In addition, Spark provides a unified platform for a variety of data processing tasks. It is a general-purpose framework that can be used for a broad range of big data applications. It is an ideal platform for interactive data analysis and applications that use iterative algorithms.

Spark's programming model is based on an abstraction called RDD. Conceptually, an RDD looks similar to a Scala collection. It represents data as a partitioned collection, which is distributed across a cluster of nodes, and provides functional methods for processing data.

CHAPTER 4

■ ■ ■

Interactive Data Analysis with Spark Shell

One of the reasons for Spark's hockey-stick growth is its usability. It not only provides a rich expressive API in multiple languages, but it also makes it easy to get started. It comes with a command-line tool called *Spark shell*, which allows you to interactively write Spark applications in Scala. The Spark shell is similar to the Scala shell, discussed in Chapter 2. In fact, it is based on the Scala shell.

The Spark shell provides a great environment for not only interactively analyzing data, but also learning Spark. You can install it on your local development machine in less than a minute and then start experimenting with Spark.

The goal of this chapter is get your hands dirty with the concepts introduced in Chapter 3. Let's use the Spark shell to play with and become familiar with Spark's programming interface.

■ **Note** The Scala and Spark shells are both command-line REPL (read-evaluate-print-loop) tools. A REPL enables interactive programming. It allows you to type an expression and have it immediately evaluated. It reads an expression, evaluates it, prints the result on the console, and waits for the next expression.

Getting Started

Getting started with Spark is simple. You can start writing Spark code in three easy steps: download, extract, and run.

Download

The Spark shell comes bundled with Spark. You can download Spark from the following web site:

```
http://spark.apache.org/downloads.html
```

The download site provides links to pre-built Spark binaries for different Hadoop versions and Hadoop distributions. It also allows you to download Spark source code and create a custom build on your machine. The easiest and fastest option is to download pre-built Spark binaries. For learning purposes, you can download Spark pre-built for any Hadoop version. Note that you do not need to install Hadoop to use Spark.

Extract

After you have downloaded the Spark binaries, unzip or untar the files on your computer. The Spark shell is available in the bin directory. For example, if you unzipped the file in a directory called SPARK_HOME, the spark-shell is available under SPARK_HOME/bin. During this chapter, SPARK_HOME will be used as a reference to the home directory for Spark on your computer. It refers to the directory where you unzipped the Spark files.

Spark runs on Linux, Mac OS, and Windows. The only prerequisite is to have Java installed on the system on which you want to run Spark. Java should be in the system path or the JAVA_HOME environment variable should point to the directory where Java is installed.

The Spark shell is verbose. It outputs a lot of information, which is useful if you want to understand the internals of Spark. You can reduce Spark's verbosity by changing the log level from INFO to WARN in the log4j.properties file in the SPARK_HOME/conf directory. Here are the steps:

1. Rename the log4j.properties.template file under the SPARK_HOME/conf to log4j.properties

2. Open the log4j.properties file in any text editor and look for this line: log4j.rootCategory=INFO, console

3. In the log4j.rootCategory line, replace INFO with WARN. After the change, the preceding line looks like this: log4j.rootCategory=WARN, console

4. Save the file.

Run

The Spark shell can be launched from a terminal in Linux or Mac. On Windows, it is recommended to launch it from a Windows PowerShell. You can use the following sequence of commands to launch the Spark shell.

```
cd $SPARK_HOME
./bin/spark-shell
```

If everything goes well, you should see a screen like the one shown in Figure 4-1.

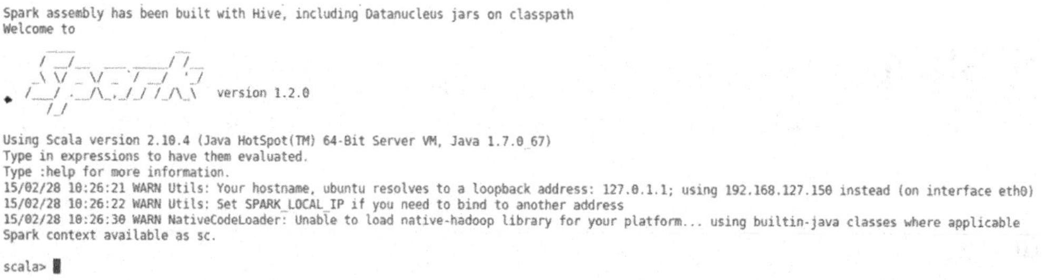

```
Spark assembly has been built with Hive, including Datanucleus jars on classpath
Welcome to
      ____              __
     / __/__  ___ _____/ /__
    _\ \/ _ \/ _ `/ __/  '_/
   /___/ .__/\_,_/_/ /_/\_\   version 1.2.0
      /_/

Using Scala version 2.10.4 (Java HotSpot(TM) 64-Bit Server VM, Java 1.7.0_67)
Type in expressions to have them evaluated.
Type :help for more information.
15/02/28 10:26:21 WARN Utils: Your hostname, ubuntu resolves to a loopback address: 127.0.1.1; using 192.168.127.150 instead (on interface eth0)
15/02/28 10:26:22 WARN Utils: Set SPARK_LOCAL_IP if you need to bind to another address
15/02/28 10:26:30 WARN NativeCodeLoader: Unable to load native-hadoop library for your platform... using builtin-java classes where applicable
Spark context available as sc.

scala> █
```

Figure 4-1. *Spark shell welcome screen*

The Spark shell creates and makes an instance of the SparkContext class available as sc. The Scala prompt indicates that the Spark shell is ready to evaluate any expression you type. Since the Spark shell is based on the Scala shell, you can enter any Scala expression after the Scala> prompt.

REPL Commands

In addition to Scala expressions, you can also enter REPL commands in the Spark shell. These are not programming code, but commands for controlling the REPL environment. The REPL commands begin with a colon. The following briefly describes common commands:

- :help

  Type :help to see a list of all the commands supported by the Spark shell.

- :quit

  Type :quit to exit from the Spark shell.

- :paste

  Type :paste to enter into paste mode. The paste mode allows you to enter or paste multi-line code blocks. It is useful when you want to copy a multi-line code block and paste it into the Spark shell. Press Ctrl-D when you are done. The Spark shell evaluates the multi-line code block after you press Ctrl-D.

Using the Spark Shell as a Scala Shell

Let's try a few simple Scala expressions to get a feel for the Spark shell.

```
scala> 2+2
res0: Int = 4
```

The first line shows the expression that you typed. After you press the Enter key, the Spark shell evaluates the expression that you typed and prints the result on the console. The second line shows the output printed by the Spark shell. It returned the result of the expression in a variable named res0, which is of type Int. You can create your own variable to store the result of an expression, as shown next.

```
scala> val sum = 2 + 2
sum: Int = 4
```

In this case, Scala evaluates the expression on the right-hand side of the equal sign and stores it in the variable sum. It then prints the type and the value of the variable sum. Note that Scala inferred the type of the variable sum from the expression that you typed.

So far, you have not really done any Spark programming. You used the Spark shell as a Scala shell. It would be good to refresh your Scala knowledge, since you will write data processing code in Scala using the Spark shell.

Number Analysis

Let's do a quick recap of some of the concepts discussed in Chapter 3. RDD is Spark's primary abstraction for representing and working with data. A Spark application processes or analyzes data using the methods defined in the RDD class. An RDD can be created from a Scala collection or a data source. An instance of the SparkContext class is required to create an RDD. As mentioned earlier in this chapter, the Spark shell automatically creates an instance of the SparkContext class and makes it available in a variable named sc.

In this section, you will use the Spark shell to analyze an RDD of numbers. You will create this RDD from a Scala collection. You will probably never create an RDD from a Scala collection in a real-world application; however, being able to create an RDD from a Scala collection makes it easy to learn Spark. The parallelize method in the SparkContext class is generally used for learning purposes. The RDD interface remains the same, irrespective of whether an RDD is created from a Scala collection, a file in HDFS, or any other Hadoop-supported storage system.

First, create a list of integers from 1 to 1000.

```
scala> val xs = (1 to 1000).toList
xs: List[Int] = List(1, 2, 3, 4, 5, 6, 7, 8, 9, 10, 11, 12, 13, 14, 15, 16, 17, 18, 19, 20,...
```

The first line shows the code that you typed after the scala> prompt. The second line shows the output printed by the Spark shell. If the result of an expression is a large collection, it prints only the first few elements.

Next, use SparkContext's parallelize method to create an RDD from the Scala list that you just created.

```
scala> val xsRdd = sc.parallelize(xs)
xsRdd: org.apache.spark.rdd.RDD[Int] = ParallelCollectionRDD[52] at parallelize at
<console>:19
```

Now filter the even numbers in the newly created RDD using Spark.

```
scala> val evenRdd = xsRdd.filter{ _ % 2 == 0}
evenRdd: org.apache.spark.rdd.RDD[Int] = FilteredRDD[53] at filter at <console>:21
```

Remember that the filter method of the RDD class is a transformation. It is a lazy operation, so Spark does not immediately apply the anonymous function that you passed to the filter method.

An action that is often used with RDDs is count. It returns the number of elements in an RDD. As discussed in the previous chapter, an action triggers the computations of the transformations.

```
scala> val count = evenRdd.count
count: Long = 500
```

Let's check the first number in the evenRdd object using the first action.

```
scala> val first = evenRdd.first
first: Int = 2
```

Next, let's check the first five numbers in the evenRdd object.

```
scala> val first5 = evenRdd.take(5)
first5: Array[Int] = Array(2, 4, 6, 8, 10)
```

See how easy it is to get started with and then experiment with the Spark API using the Spark shell?

Log Analysis

Let's now use the Spark shell to analyze data in a file. You will analyze an event log file generated by an application. For learning purposes, you will use a pseudo cluster of one machine—your laptop. To keep things simple, you will use a file on your local file system as the data source. It is a simple setup to get started; however, the code that you create in this section will work exactly the same way on a cluster with hundreds of nodes and a distributed data source. That is the beauty of the Spark API: a Spark application can easily scale from processing kilobytes of data on a single laptop to processing petabytes of data on a large cluster without any code change.

For simplicity sake, you will analyze a fake log file in this section. The event logs are stored in a space-separated text format. The first column represents the date and time when an event was logged. Instead of real timestamps, it stores an incrementing number sequence. A real log file will have date and timestamps in this column. The second column stores the severity of an event. It can be DEBUG, INFO, WARN, or ERROR. The third column stores the name of the module or component that generated the log. The last column stores the description of an event. This is a typical format of a log file, but there are a number of variations of this format. For example, some developers choose a different order for the columns. Sometimes people also log information about the thread that was executing the application code when an event occurred.

The first step in analyzing a dataset using Spark is to create an RDD from a data source. Similar to creating an RDD from a Scala collection, an instance of the SparkContext class is required to create an RDD from a data source.

Let's assume that the fake event log file is stored under the data directory. You can read it using its relative path.

```scala
scala> val rawLogs = sc.textFile("data/app.log")
rawLogs: org.apache.spark.rdd.RDD[String] = data/app.log MappedRDD[55] at textFile at <console>:17
```

Make sure that the file path is valid and that the file exists on your machine. If you have a different directory structure or you launched the Spark shell from a different directory, make appropriate changes to the code.

The output printed by the Spark shell in the second line shows that Spark has created an RDD of Strings. Each line in the event log file is stored as a String in the rawLogs RDD.

Remember that Spark does not yet actually read the app.log file at this point. The textFile method is a lazy operation. Spark just remembers how to create the rawLogs RDD. It will read the file when an action is called on the rawLogs RDD or an RDD created directly or indirectly from rawLogs. This means that even if you specify a wrong path or filename to the textFile method, Spark will not throw an exception right away.

If you wanted to analyze an event log file stored on HDFS, you would use the HDFS path URI to create an RDD from that file.

```scala
scala> val rawLogs = sc.textFile("hdfs://...")
```

The textFile method takes an optional second input parameter, which specifies the minimum number of partitions in the returned RDD. As discussed in Chapter 3, an RDD is a partitioned collection of elements. For an HDFS file, Spark creates a partition for each HDFS file block. You can request a higher number of partitions. The parallelism within a job depends on the number of RDD partitions; a higher number of partitions increases parallelism.

Let's convert each log line to lowercase so that you don't have to worry whether a word is in uppercase or lowercase in the input file. You will also trim each line to remove leading and trailing whitespaces. The map method can be used to transform the original logs to trimmed logs in lowercase.

```scala
scala> val logs = rawLogs.map {line => line.trim.toLowerCase()}
logs: org.apache.spark.rdd.RDD[String] = MappedRDD[56] at map at <console>:19
```

Since the event log data will be queried more than once, let's cache the logs RDD.

```
scala> logs.persist()
res18: logs.type = MappedRDD[56] at map at <console>:19
```

Remember that Spark does not immediately materialize or cache an RDD in memory when you call the persist method. In fact, it has not yet even read the log file from the disk yet. All of that happens after you call the first action method.

Let's make it happen by calling the count action on the logs RDD.

```
scala> val totalCount = logs.count()
totalCount: Long = 103
```

If the file path specified to the textFile method were incorrect, you will get an exception because Spark cannot find the file.

Let's filter the logs since you are interested only in logs with error severity.

```
scala> val errorLogs = logs.filter{ line  =>
                              val words = line.split(" ")
                              val logLevel = words(1)
                              logLevel == "error"
                       }
errorLogs: org.apache.spark.rdd.RDD[String] = FilteredRDD[57] at filter at <console>:21
```

This code first splits a line into an array of words. Since you know that the second column stores the severity of an event, you fetch the second element from the array and check if it is equal to "error". The filter method of an RDD keeps only those elements for which the anonymous function returns true.

A concise version of the same code is shown here.

```
scala> val errorLogs = logs.filter{_.split(" ")(1) == " error"}
errorLogs: org.apache.spark.rdd.RDD[String] = FilteredRDD[4] at filter at <console>:16
```

Since you created the log file, you know that columns are space-separated. If you are not sure whether the columns in a file are separated by a single space, multiple spaces, or tabs, you can use a regex to split the columns.

```
scala> val errorLogs = logs.filter{_.split("\\s+")(1) == "error"}
errorLogs: org.apache.spark.rdd.RDD[String] = FilteredRDD[5] at filter at <console>:16
```

Now let's check the number of logs that have error severity.

```
scala> val errorCount = errorLogs.count()
errorCount: Long = 26
```

When you troubleshoot a problem, you want to see the first error, which you can fetch using the action method first.

```
scala> val firstError = errorLogs.first()
firstError: String = 4 error module1 this is an error log
```

If you want to see the first three errors, you can use the Spark take action method.

```scala
scala> val first3Errors = errorLogs.take(3)
first3Errors: Array[String] = Array(4 error module1 this is an error log, 8 error module2
this is an error log, 12 error module3 this is an error log)
```

Suppose that for some reason you want to find the longest log line in the log file. You first transform the logs RDD to another RDD that contains the length of each log line. Then you use the reduce action method to find the max value in the new RDD.

```scala
scala> val lengths = logs.map{line => line.size}
lengths: org.apache.spark.rdd.RDD[Int] = MappedRDD[6] at map at <console>:16

scala> val maxLen = lengths.reduce{ (a, b) => if (a > b) a else b }
maxLen: Int = 117
```

If you want to find the logs with most words, you can use the following code.

```scala
scala> val wordCounts = logs map {line => line.split("""\s+""").size}
wordCounts: org.apache.spark.rdd.RDD[Int] = MappedRDD[7] at map at <console>:16

scala> val maxWords = wordCounts reduce{ (a, b) => if (a > b) a else b }
maxWords: Int = 20
```

This code transforms the logs RDD to another RDD that has the number of words in each line. To achieve this, it splits each line into an array of words and gets the size of the array. The next step is same as in the previous example. It uses the reduce method to the find the max element in the new RDD.

If you want to save the error logs in a file for further processing by another application, you can use the saveAsTextFile method.

```scala
scala> errorLogs.saveAsTextFile("data/error_logs")
```

Note that the input parameter to the saveAsTextFile method is a directory name. For each RDD partition, Spark creates a file under that directory.

So far, you have worked with single-line expressions. Let's write a multi-line function. You will create a function that takes an event log as input and returns its severity. Although the Spark shell allows to you to write multi-line line code blocks, it is not an ideal environment for this. You can write your function in any text editor and copy it into the Spark shell. To copy and paste a multi-line code block into the Spark shell, you use the paste command.

```scala
scala> :paste
// Entering paste mode (ctrl-D to finish)

def severity(log: String): String = {
    val columns = log.split("\\s+", 3)
    columns(1)
  }

// Exiting paste mode, now interpreting.

severity: (log: String)String
```

The severity function splits an event log into three words and returns the second word, which is the severity of the log. Note that arrays in Scala have a zero-based index.

You now use this function to calculate the count of logs by severity.

```scala
scala> val pairs = logs.map { log => (severity(log), 1)}
pairs: org.apache.spark.rdd.RDD[(String, Int)] = MappedRDD[3] at map at <console>:18

scala> val countBySeverityRdd = pairs.reduceByKey{(x,y) => x + y}
countBySeverityRdd: org.apache.spark.rdd.RDD[(String, Int)] = MapPartitionsRDD[6] at
reduceByKey at <console>:25
```

The preceding code first transforms the logs RDD to an RDD of key-value pairs, where the key is the severity of a log and the value is 1. Next, you use the reduceByKey method to count the number of elements with the same key. Note that the reduceByKey method returns an RDD of key-value pairs. The returned RDD has one entry for each unique value of log severity. You can use the collect method to fetch the values from this RDD.

```scala
scala> val countBySeverity= countBySeverityRdd.collect()
countBySeverity: Array[(String, Int)] = Array((warn,25), (info,27), (error,26), (debug,25))
```

You can also save it to disk.

```scala
scala> countBySeverityRdd.saveAsTextFile("data/log-counts-text")
```

As previously mentioned, the input parameter to the saveAsTextFile method is a directory name. Spark creates a file for each RDD partition in the specified directory.

Since countBySeverityRdd is an RDD of key-values pairs, you can also save it in SequenceFiles.

```scala
scala> countBySeverityRdd.saveAsSequenceFile("data/log-counts-seq")
```

A slightly more concise solution for counting the number of logs by severity is shown next.

```scala
scala> val countBySeverityMap = pairs.countByKey()
```

This example uses countByKey instead of the reduceByKey method. The latter is a generic method that can be used for not only calculating count but also sum, product, or any custom aggregation of the values for each key. In fact, countyByKey internally uses reduceByKey and collect.

Summary

The Spark shell is a command-line tool that allows you to interactively write Spark code. It makes it easy to get started with and learn Spark. It is also a powerful tool for conducting interactive data analysis with Spark.

In this chapter, you used the concepts introduced in the previous chapter for two different tasks. First, you created an RDD from a Scala collection and analyzed it with Spark. Next, you analyzed an event log file using Spark.

The next chapter covers developing a complete Spark application. You will learn how to write, compile, and deploy a Spark application on a cluster.

CHAPTER 5

■ ■ ■

Writing a Spark Application

This chapter discusses how to write a data processing application in Scala using Spark. Chapter 2 covered the basics of Scala, which you need to know to get started.

The mechanics of writing a Spark application in Scala is not very different from writing a standard Scala application. A Spark application is basically a Scala application that uses Spark as a library. The API provided by the Spark library was discussed in Chapter 3. You will use that API to write a simple data processing application. The goal is to show you how to write, compile, and run a Spark application, so the data processing logic will be kept simple.

Hello World in Spark

It is customary to write a "Hello World!" program when one learns a new programming language. The "Hello World!" equivalent of a big data application is the word-count application. It counts the number of times each unique word occurs in a document.

Let's write the big data "Hello World!" application using Spark. You will call it WordCount. You can write this code in any text editor, Scala IDE, or IntelliJ IDEA.

```scala
import org.apache.spark.SparkContext
import org.apache.spark.SparkContext._

object WordCount {
  def main(args: Array[String]): Unit = {
    val inputPath = args(0)
    val outputPath = args(1)
    val sc = new SparkContext()
    val lines = sc.textFile(inputPath)
    val wordCounts = lines.flatMap {line => line.split(" ")}
                        .map(word => (word, 1))
                        .reduceByKey(_ + _)
    wordCounts.saveAsTextFile(outputPath)
  }
}
```

WordCount is a Scala application that uses Spark as a library. It takes two input parameters or arguments. The first argument is the path to the input dataset. WordCount counts the number of occurrences of each unique word in the dataset located at this path. The second parameter is the path for storing the results.

You can use this simple application as a starter template for writing complex Spark applications. Note that the same code will work on both a small dataset on a local file system and a large dataset (multi-terabytes) stored on a distributed file system such as HDFS.

Let's walk through the code, line by line, to understand the basic structure of a Spark application.

```
import org.apache.spark.SparkContext
```

First, you import the `SparkContext` class since you will be creating an instance of this class. Note that the Spark shell creates and makes available an instance of the `SparkContext` class, so you do not need to import the `SparkContext` class when using the Spark shell.

```
import org.apache.spark.SparkContext._
```

This line imports all the definitions from the `SparkContext` object. The `SparkContext` object includes a number of implicit functions for implicit conversions, which eliminate boilerplate code and simplify code. These implicit functions make it easy to use various Spark features.

```
object WordCount {
  def main(args: Array[String]): Unit = {
    ...
  }
}
```

Next, you define the `WordCount` object. A Spark application is required to define a singleton object with a method named `main` that takes an array of `String` as an input argument. You can name the singleton object in your application whatever you want, but it should have a `main` method with the same signature, as shown earlier.

The `main` method is a Spark application's entry point. This is where the execution of a Spark application begins when you run it. The command-line arguments provided by a user running an application are passed to the `main` method through the `args` argument. The `args` parameter is an array of `String`. The first element in the array contains the first argument; the second element contains the second argument, and so on. The `main` method does not return any value.

```
val inputPath = args(0)
```

This line fetches the first command-line argument passed to the application. The first argument should be the path to the input dataset. The path can be a complete or relative path to a file on a local file system, HDFS, or S3.

For readability sake, I have not included any error-checking code inside the `main` method. If a user forgets to provide arguments to the application, it will crash with an exception. In a real-world application, you would first check whether a user launched your application with the expected number of arguments of the right types.

```
val outputPath = args(1)
```

Next, you fetch the destination path for storing the results of the computation. Note that the destination path should be the name of a directory. Spark creates a file for each data partition in this directory. The output destination can be a directory on the local file system, HDFS, or S3.

```
val sc = new SparkContext()
```

Next, you create an instance of the SparkContext class, which is the main entry point into the Spark library. It is required for creating other Spark objects, including RDDs. A Spark application needs to create an instance of the SparkContext class to access the functionality provided by Spark.

```
val lines = sc.textFile(inputPath)
```

The preceding code creates an RDD from the input dataset. The textFile method creates an RDD of Strings. Each line of text in the input dataset is stored as an element of type String in the lines RDD, which has the same number of elements as the number of lines in the input dataset.

```
val wordCounts = lines.flatMap{line => line.split(" ")}
                      .map{word => (word, 1)}
                      .reduceByKey{(x,y) => x + y}
```

The core data processing logic for the application is implemented by the preceding code. It uses a bunch of RDD transformations to calculate the count of each unique word in the input dataset. The transformations used in the code are functional transformations. You call higher-order RDD methods and pass functions that specify the exact transformations that you want to run on each data element.

First, you use the flatMap method to generate an RDD of words from an RDD of lines of text. Each element in the lines RDD is a String representing a line of text in the input dataset. You use the String class' split method to break a line of text into words. For simplicity sake, you assume that the words are separated by a single space. You can handle complex cases by passing a regex to the split method.

Next, you use the map method to transform an RDD of words to an RDD of key-value pairs, where a key is a word from the input dataset and the value is always 1.

Next, you use the reduceByKey method to calculate the count of each unique word in the input dataset. The reduceByKey method is available only for the RDD of key-value pairs. This is the reason you created an RDD of key-value pairs using the map method in the previous step.

```
wordCounts.saveAsTextFile(outputPath)
```

Finally, you use the saveAsTextFile method to save the RDD returned by the reduceByKey method to a directory specified by a user of the application. This step completes the application.

Compiling and Running the Application

To run the WordCount application, you first compile its source code and create a jar file. Next, you use the spark-submit script that comes with Spark to run the application.

The tool commonly used to build a Spark application written in Scala is called *sbt*. Although most developers use either Scala IDE or IntelliJ IDEA for developing Scala applications, you can write a Scala application in any text editor and use sbt to compile and run it. You can also use Maven.

sbt (Simple Build Tool)

sbt is an open source interactive build tool. You can use it to compile and run a Scala application. You can also use it to run your tests. sbt is a powerful build tool; mastering it takes some time, but you can get quickly started after you have learned the basics.

sbt can be downloaded from www.scala-sbt.org/download.html. I recommend reading the "Getting started with sbt" section in the sbt tutorial available at www.scala-sbt.org/documentation.html. It provides instructions for installing sbt on different operating systems. It also covers the basic concepts you need to know to use sbt.

Build Definition File

sbt requires a build definition file for a Scala project. A build definition file defines project settings for sbt. For the WordCount application, save the build definitions in a file named `wordcount.sbt`. The contents of the `wordcount.sbt` file are shown next.

```
name := "word-count"

version := "1.0.0"

scalaVersion := "2.10.6"

libraryDependencies += "org.apache.spark" %% "spark-core" % "1.5.1" % "provided"
```

Each line in the `wordcount.sbt` file defines a setting for the WordCount application. Note that a blank line between two settings is mandatory. Future versions of sbt may remove this requirement.

The first line in the `wordcount.sbt` file assigns a name to the application. The third line assigns a version number to the application. sbt uses the name and version string in naming the jar file that it generates. The fifth line specifies the version of Scala that sbt should use to compile the application. The last line specifies Spark core as a third-party library dependency for WordCount. It also specifies the version of the Spark core library. sbt downloads not only the Spark core jar file, but also all of its dependencies.

Note that the version specified for the Spark core library should match with the version of the Spark binaries that you downloaded. The Spark developer community releases a new version of Spark every three months. At the time of writing this chapter, the current version of Spark is 1.5.1, but a newer version may be available for download by the time you read this. In other words, replace `1.5.1` with the version number that matches with the version of Spark that you downloaded.

Directory Structure

For a simple application, you can put both the source code and the sbt build definition files in the same directory. For the WordCount application, create a directory with the same name and save the source code in a file named `WordCount.scala`.

For complex projects, please refer to the sbt documentation for the directory structure convention. As previously mentioned, a sbt tutorial is available at `www.scala-sbt.org/documentation.html`.

Compiling the Code

Let's now compile the source code for the WordCount application and create a jar file using sbt. First, go to the `WordCount` directory and then run sbt, as shown next.

```
$ cd WordCount
$ sbt package
```

sbt creates a jar file named `word-count_2.10-1.0.0.jar` under `./target/scala-2.10`.

Running the Application

A Spark application can be launched using the `spark-submit` script that comes with Spark. It is available in the `SPARK_HOME/bin` directory.

The `spark-submit` script can be used for both running a Spark application on your development machine and deploying it on a real Spark cluster. It provides a unified interface for running a Spark application with any Spark-supported cluster manager.

Let's run the WordCount application in local mode using the standalone cluster manager that comes pre-packaged with Spark.

```
$ ~/path/to/SPARK_HOME/bin/spark-submit --class "WordCount" --master local[*] \
                  target/scala-2.10/word-count_2.10-1.0.0.jar \
                  path/to/input-files \
                  path/to/output-directory
```

The `spark-submit` script supports a number of options, which can be specified using the following notation.

```
--option_name option_value
```

After the options, specify the path to the application's jar file followed by the application arguments. In the preceding example, `target/scala-2.10/word-count_2.10-1.0.0.jar` is the relative path to the application jar file.

The commonly used options with spark-submit are briefly described next.

--master MASTER_URL

The `master` option is used to specify the master URL for the Spark cluster. The master URL tells the `spark-submit` script where you want to run your Spark application.

If you want to run Spark locally with one worker thread, you can use `local` as the master URL. In local mode, you can specify the number of worker threads using this notation:

```
--master local[N]
```

N stands for the number of worker threads. Replace N with * if you want Spark to use as many worker threads as logical cores on your machine.

```
--master local[*]
```

If you want to run your application on a Spark standalone cluster, you need to specify the host and port of the Spark standalone cluster master.

```
--master spark://HOST:PORT
```

--class CLASS_NAME

The `class` option specifies the class that contains the `main` method for your Spark application. It is the entry point for your application.

--jars JARS

The jars option can be used to specify a comma-separated list of local jar files to include on the driver and executor classpaths. This option is useful if your application uses third-party libraries that are not included in your application jar file. Spark sends the specified jars to the executors running on the worker nodes.

You can get the complete list of options for the spark-submit script using the --help option, as shown next.

```
$ ~/path/to/ SPARK-HOME /bin/spark-submit --help
Usage: spark-submit [options] <app jar | python file> [app options]
Options:
  --master MASTER_URL         spark://host:port, mesos://host:port, yarn, or local.

  --deploy-mode DEPLOY_MODE   Whether to launch the driver program locally ("client") or
                              on one of the worker machines inside the cluster ("cluster")
                              (Default: client).

  --class CLASS_NAME          Your application's main class (for Java / Scala apps).

  --name NAME                 A name of your application.

  --jars JARS                 Comma-separated list of local jars to include on the driver
                              and executor classpaths.

  --py-files PY_FILES         Comma-separated list of .zip, .egg, or .py files to place
                              on the PYTHONPATH for Python apps.

  --files FILES               Comma-separated list of files to be placed in the working
                              directory of each executor.

  --conf PROP=VALUE           Arbitrary Spark configuration property.

  --properties-file FILE      Path to a file from which to load extra properties. If not
                              specified, this will look for conf/spark-defaults.conf.

  --driver-memory MEM         Memory for driver (e.g. 1000M, 2G) (Default: 512M).

  --driver-java-options       Extra Java options to pass to the driver.

  --driver-library-path       Extra library path entries to pass to the driver.

  --driver-class-path         Extra class path entries to pass to the driver. Note that
                              jars added with --jars are automatically included in the
                              classpath.

  --executor-memory MEM       Memory per executor (e.g. 1000M, 2G) (Default: 1G).

  --help, -h                  Show this help message and exit
  --verbose, -v               Print additional debug output
```

```
Spark standalone with cluster deploy mode only:
  --driver-cores NUM        Cores for driver (Default: 1).
  --supervise               If given, restarts the driver on failure.

Spark standalone and Mesos only:
  --total-executor-cores NUM  Total cores for all executors.

YARN-only:
  --executor-cores NUM      Number of cores per executor (Default: 1).
  --queue QUEUE_NAME        The YARN queue to submit to (Default: "default").
  --num-executors NUM       Number of executors to launch (Default: 2).
  --archives ARCHIVES       Comma separated list of archives to be extracted into the
                            working directory of each executor.
```

Monitoring the Application

Spark comes pre-packaged with a web-based application for monitoring an application running on a Spark cluster. You can monitor a Spark application from any web browser. You just need the IP address or hostname of the machine running the driver program and the monitoring UI port.

The default application monitoring port is 4040; however, if that port is not available, the web UI tries port 4041. If port 4041 is also not available, it will try port 4042 and so on.

The web-based monitoring application provides useful information about an application running on a Spark cluster. Chapter 11 covers the monitoring UI in more detail.

Debugging the Application

Debugging a distributed application can be a difficult task. Standard debugging tools designed for applications running on a single machine are not of much help.

One useful tool for troubleshooting performance-related issues is the web-based monitoring application provided by Spark. It provides detailed information about all the jobs created by an application. It also shows the stages and tasks for each job. Careful monitoring of an application's jobs, stages, and tasks can provide clues to performance issues.

Another useful debugging technique is the use of logs. A good practice is to log events. By generating logs at strategic places in your application, you can monitor the runtime behavior of your application.

Spark generates tremendous amount of logs, which provide insight into the internal workings of Spark. You can configure both the destination for the logs and the amount of logs generated by Spark through the log4j.properties file under the SPARK_HOME/conf directory. By default, on a standalone Spark cluster, the log files for the Spark master are generated under the SPARK_HOME/logs directory. The executor logs on the worker nodes are available under the SPARK_HOME/work directory.

There are many logging frameworks that you can use for logging events in your own application. The two popular ones for Java and Scala applications are Log4j (logging.apache.org/log4j) and Logback (logback.qos.ch). These libraries make it really easy to logs events in a flexible manner.

Summary

A Spark application written in Scala is not very different from a standard Scala application. It is basically a Scala application that uses Spark as a library. It uses the Spark API for distributed data processing on a cluster of computers.

Developing and deploying a Spark application is a three-step process. First, you write a Spark application using any text editor, Scala IDE, or IntelliJ IDEA. Next, build it with sbt or Maven. Finally, deploy it using the `spark-submit` script.

As an application developer, you focus only on the data processing logic in your code and Spark takes care of distributing your data processing code across a cluster of nodes. It handles all the messy details of distributed computing and fault handling. The same code works on both a single machine and a cluster of thousand machines. Without changing a single line of code, you can scale your application to process anywhere from a few bytes to petabytes of data.

The next chapter covers live data stream processing with Spark. It will discuss a Spark library specifically designed for data stream processing.

CHAPTER 6

■ ■ ■

Spark Streaming

Batch processing of historical data was one of the first use cases for big data technologies such as Hadoop and Spark. In batch processing, data is collected for a period of time and processed in batches. A batch processing system processes data spanning from hours to years, depending on the requirements. For example, some organizations run nightly batch processing jobs, which process data collected throughout the day by various systems.

Batch processing systems have high latency. Depending on the volume of data, it may take anywhere from a few minutes to a few hours to process a batch. Some organizations run nightly batch processing jobs that run for 6 to 12 hours on a cluster of hundreds of machines. Thus, there is a long wait before you can see the results generated by a batch processing application. In addition, since data is not immediate processed, the gap between the time when data is collected and the time when the batch processing result becomes available is even longer. This time gap is acceptable for a certain class of applications.

However, sometimes data needs to be processed and analyzed as it is collected. For example, fraud detection in an e-commerce system must happen in real time. Similarly, network intrusion or security breach detection must be in real time. Another example is application or device failure detection in a data center. To prevent a long downtime, data must be processed right away.

One of the challenges with live data stream processing is handling high-velocity data in real time or near real time. A data stream processing application running on a single machine will not be able to handle high-velocity data. A distributed stream processing framework addresses this issue.

This chapter introduces Spark Streaming. The introduction is followed by a detailed discussion of the application programming interface provided by Spark Streaming. At the end of the chapter, you will develop an application using Spark Streaming.

Introducing Spark Streaming

Spark Streaming is a distributed data stream processing framework. It makes it easy to develop distributed applications for processing live data streams in near real time. It not only provides a simple programming model but also enables an application to process high-velocity stream data. It also allows the combining of data streams and historical data for processing.

Spark Streaming Is a Spark Add-on

Spark Streaming is a Spark library that runs on top of Spark. It extends Spark for data stream processing. It provides higher-level abstractions for processing streaming data, but under the hood, it uses Spark (see Figure 6-1).

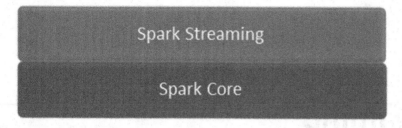

Figure 6-1. *Spark Streaming runs on top of Spark core*

Since Spark Streaming runs on top of Spark, it provides a scalable, fault-tolerant, and high-throughput distributed stream processing platform. It inherits all the features and benefits of Spark core. The processing capability of Spark Streaming application can be easily increased by adding more nodes to a Spark cluster.

In addition, Spark Streaming can be used along with other Spark libraries, such as Spark SQL, MLlib, Spark ML, and GraphX. A data stream can be analyzed using SQL. Machine learning algorithms can be applied to a data stream. Similarly, graph processing algorithms can be applied to stream data. Thus, Spark Streaming makes the power of the complete Spark stack available for processing data streams.

High-Level Architecture

Spark Streaming processes a data stream in micro-batches. It splits a data stream into batches of very small fixed-sized time intervals. Data in each micro-batch is stored as an RDD, which is then processed using Spark core (see Figure 6-2). Any RDD operation can be applied to an RDD created by Spark Streaming. The results of the RDD operations are streamed out in batches.

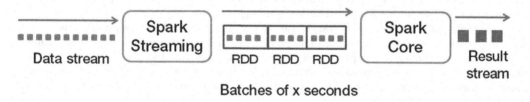

Figure 6-2. *Stream to micro-batches*

Data Stream Sources

Spark Streaming supports a variety of data stream sources, including TCP socket, Twitter, Kafka, Flume, Kinesis, ZeroMQ, and MQTT. It can also be used to process a file as a stream. In addition, you can extend it to process data from a custom streaming data source.

The streaming data sources for which Spark Streaming has built-in support can be grouped into two categories: basic sources and advanced sources.

- **Basic data stream sources** include TCP sockets, Akka Actors, and files. Spark Streaming includes the libraries required to process data from these sources. A Spark Streaming application that wants to process data streams from a basic source needs to link only against the Spark Streaming library.

- **Advanced data stream sources** include Kafka, Flume, Kinesis, MQTT, ZeroMQ, and Twitter. The libraries required for processing data streams from these sources are not included with Spark Streaming, but are available as external libraries. A Spark Streaming application that wants to process stream data from an advanced source must link against not only the Spark Streaming library, but also the external library for that source.

Receiver

A Receiver receives data from a streaming data source and stores it in memory. Spark Streaming creates and runs a Receiver on a worker node for each data stream. An application can connect to multiple data streams to process data streams in parallel.

Destinations

The results obtained from processing a data stream can be used in a few different ways (see Figure 6-3). The results may be fed to another application, which may take some action or just display it. For example, a Spark Streaming application may feed the results to a dashboard application that is updated continuously. Similarly, in a fraud detection application, the results may trigger cancellation of a transaction. The results can also be stored in a storage system such as a file or a database.

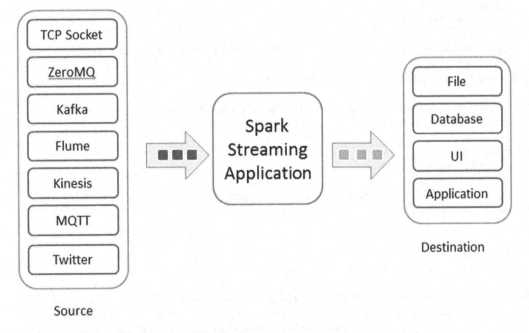

Figure 6-3. *Stream data sources and destinations*

Application Programming Interface (API)

The Spark Streaming library is written in Scala, but it provides an application programming interface (API) in multiple languages. At the time this book was written, the Spark Streaming API was available in Scala, Java, and Python.

The Spark Streaming API consists of two key abstractions, StreamingContext and Discretized Streams. A Spark Streaming application processes a data stream using these two abstractions.

This section covers StreamingContext and Discretized Streams in detail. It discusses the various operations supported by these abstractions and how you can create them. It also discusses the basic structure of a Spark Streaming application.

StreamingContext

StreamingContext, a class defined in the Spark Streaming library, is the main entry point into the Spark Streaming library. It allows a Spark Streaming application to connect to a Spark cluster. It also provides methods for creating an instance of the data stream abstraction provided by Spark Streaming.

Every Spark Streaming application must create an instance of this class.

Creating an Instance of StreamingContext

Creating an instance of StreamingContext is similar to creating an instance of SparkContext. It can be created using the same parameters that are required to create an instance of the SparkContext class. However, it takes an additional argument, which specifies the time interval for splitting a data stream into micro-batches.

```
import org.apache.spark._
import org.apache.spark.streaming._

val config = new SparkConf().setMaster("spark://host:port").setAppName("big streaming app")
val batchInterval = 10
val ssc = new StreamingContext(conf, Seconds(batchInterval))
```

Alternatively, if you already have an instance of the SparkContext class, you can use it to create an instance of the StreamingContext class.

```
import org.apache.spark._
import org.apache.spark.streaming._

val config = new SparkConf().setMaster("spark://host:port").setAppName("big streaming app")
val sc = new SparkContext(conf)
...
...
val batchInterval = 10
val ssc = new StreamingContext(sc, Seconds(batchInterval))
```

The second argument to the StreamingContext constructor specifies the size of a micro-batch using time as a unit. A data stream is split into batches of this time duration and each batch is processed as an RDD. The preceding example specifies a batch duration of 10 seconds. Spark Streaming creates a new RDD from a streaming source every 10 seconds.

The batch size can be as small as 500 milliseconds. The upper bound for the batch size is determined by the latency requirements of your application and the available memory. The executors created for a Spark Streaming application must have sufficient memory to store the received data in memory for good performance.

You will use the variable ssc in other examples in this chapter. Instead of repeating the preceding code snippet in every example, assume that the ssc variable is defined at the beginning of the program, as shown earlier.

The StreamingContext class defines methods for creating instances of classes representing different types of data streams. They are discussed in detail in later sections. The other commonly used methods from the StreamingContext class are briefly described next.

Starting Stream Computation

The start method begins stream computation. Nothing really happens in a Spark Streaming application until the start method is called on an instance of the StreamingContext class. A Spark Streaming application begins receiving data after it calls the start method.

```
ssc.start()
```

Checkpointing

The checkpoint method defined in the StreamingContext class tells Spark Streaming to periodically checkpoint data. It takes the name of a directory as an argument. For a production application, the checkpoint directory should be on a fault-tolerant storage system such as HDFS.

```
ssc.checkpoint("path-to-checkpoint-directory")
```

A Spark Streaming application must call this method if it needs to recover from driver failures or if it performs stateful transformations. The data processed by a Spark Streaming application is conceptually a never ending sequence of continuous data. If the machine running the driver program crashes after some data has been received but before it has been processed, there is a potential for data loss. Ideally, a Spark Streaming application should be able to recover from failures without losing data. To enable this functionality, Spark Streaming requires an application to checkpoint metadata.

In addition, data checkpointing is required when an application performs stateful transformation on a data stream. A stateful transformation is an operation that combines data across multiple batches in a data stream. An RDD generated by a stateful transformation depends on the previous batches in a data stream. Therefore, the dependency tree for an RDD generated by a stateful transformation grows with time. In case of a failure, Spark Streaming reconstructs an RDD using its dependency tree. As the dependency tree grows, the recovery time increases. To prevent recovery time from becoming too high, Spark Streaming checkpoints intermediate RDDs of a stateful transformation. Therefore, Spark Streaming requires an application to call the checkpoint method prior to using a stateful transformation. The stateful transformations supported by Spark Streaming are discussed later in this chapter.

Stopping Stream Computation

The stop method, as the name implies, stops stream computation. By default, it also stops SparkContext. This method takes an optional parameter that can be used to stop only the StreamingContext, so that the SparkContext can be used to create another instance of StreamingContext.

```
ssc.stop(true)
```

Waiting for Stream Computation to Finish

The awaitTermination method in the StreamingContext class makes an application thread wait for stream computation to stop. It's syntax is

```
ssc.awaitTermination()
```

The awaitTermination method is required if a driver application is multi-threaded and the start method was called not from the main application thread but by another thread. The start method in the StreamingContext class is blocking method; it does not return until stream computation is finished or stopped. In a single-threaded driver, the main thread will wait until the start method returns. However, if the start method was called from another thread, you can prevent your main thread from exiting prematurely by calling awaitTermination.

Basic Structure of a Spark Streaming Application

Let's create an outline of a Spark Streaming application using the classes and methods discussed so far. It shows the basic structure and does not have any processing logic yet. As you progress through the chapter, you will see code snippets for different types of stream processing. You can add those snippets to this skeleton to get a complete working Spark Streaming application.

```
import org.apache.spark._
import org.apache.spark.streaming._

object StreamProcessingApp {
  def main(args: Array[String]): Unit = {
      val interval = args(0).toInt
      val conf = new SparkConf()
      val ssc = new StreamingContext(conf, Seconds(interval))

      // add your application specific data stream processing logic here
      ...
      ...
      ...

      ssc.start()
      ssc.awaitTermination()
  }
}
```

Discretized Stream (DStream)

Discretized Stream (DStream) is the primary abstraction provided by Spark Streaming for working with data streams. It represents a data stream and defines the operations that you can perform on a data stream.

DStream is defined as an abstract class in the Spark Streaming library. It defines an interface for processing a data stream. Spark Streaming provides concrete classes that implement the DStream interface for stream data from a variety of sources. I use the term DStream generically to refer to both the abstract class and the classes implementing the interface defined by the DStream class.

Spark Streaming implements a DStream as a sequence of RDDs (see Figure 6-4). It translates DStream operations to operations on the underlying RDDs.

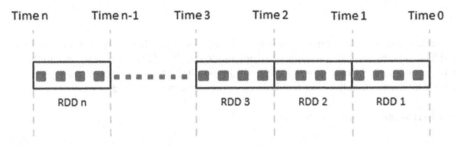

Figure 6-4. *DStream is a never-ending sequence of RDDs*

Since a DStream is a sequence of RDDs, it inherits the key RDD properties. It is immutable, partitioned, and fault tolerant.

Creating a DStream

A DStream can be created from a streaming data source or from an existing DStream by applying a transformation. Since DStream is an abstract class, you cannot directly create an instance of the DStream class. The Spark Streaming library provides factory methods for creating instances of the concrete classes that implement the interface defined by DStream.

Basic Sources

The factory methods for creating a DStream from basic sources are built-in. A Spark Streaming application needs to link only against the Spark Streaming library to use these methods. The StreamingContext class defines the methods for creating a DStream from a basic source. The commonly used methods for creating a DStream from a basic source are briefly described next.

socketTextStream

The socketTextStream method creates a DStream that represents stream data received over a TCP socket connection. It takes three input parameters. The first argument is the hostname of the data source. The second argument is the port to connect to for receiving data. The third argument, which is optional, specifies the storage level for the received data.

```
val lines = ssc.socketTextStream("localhost", 9999)
```

The default storage level is `StorageLevel.MEMORY_AND_DISK_SER_2`, which stores the received data first in memory and spills to disk if the available memory is insufficient to store all received data. In addition, it deserializes the received data and reserializes it using Spark's serialization format. Thus, this storage level incurs the overhead of data serialization, but it reduces JVM garbage collection-related issues. The received data is also replicated for fault tolerance.

You can change the storage level for the received data by explicitly providing the third argument. For example, if the batch interval of your Spark Streaming application is only a few seconds, you can improve application performance by using the `StorageLevel.MEMORY_ONLY` storage level.

```
val lines = ssc.socketTextStream("localhost", 9999, StorageLevel.MEMORY_ONLY)
```

textFileStream

The `textFileStream` method creates a DStream that monitors a Hadoop-compatible file system for new files and reads them as text files. It takes as input the name of a directory to monitor. Files must be written to the monitored directory by moving them from another location within the same file system. For example, on a Linux system, files should be written into the monitored directory using the `mv` command.

```
val lines = ssc.textFileStream("input_directory")
```

actorStream

The `actorStream` method creates a DStream with a user-implemented Akka actor Receiver.

Advanced Sources

The factory methods for creating a DStream from advanced sources such as Kafka, Flume, or Twitter are not built-in, but available through extra utility classes. To process a data stream from an advanced source, an application needs to perform the following steps:

1. Import the utility class for that source and create a `DStream` using the factory method provided by that class.

2. Link against the library that contains the utility class for that source.

3. Create an uber JAR that includes all application dependencies and deploy the application on a Spark cluster.

For example, to process tweets from Twitter, an application must import the `TwitterUtils` class and use its `createStream` method to create a DStream that receives tweets from Twitter.

```
import org.apache.spark.streaming.twitter._
...
...
val tweets = TwitterUtils.createStream(ssc, None)
```

Processing a Data Stream

An application processes a data stream using the methods defined in the DStream and related classes. DStream supports two types of operations: transformation and output operation. The transformations can be further classified into basic, aggregation, key-value, and special transformation.

Similar to RDD transformations, DStream transformations are lazily computed. No computation takes places immediately when a transformation operation is called. An output operation triggers the execution of DStream transformation operations. In the absence of an output operation on a DStream, Spark Streaming will not do any processing, even if transformations are called on that DStream.

Basic Transformation

A transformation applies a user-defined function to each element in a DStream and returns a new DStream. DStream transformations are similar to RDD transformations. In fact, Spark Streaming converts a DStream transformation method call to a transformation method call on the underlying RDDs. Spark core computes the RDD transformations.

The commonly used DStream transformations are briefly described next.

map

The map method takes a function as argument and applies it to each element in the source DStream to create a new DStream. It returns a new DStream.

For example, assume that you have an application running on local host and sending streams of text lines on TCP port 9999. The following code snippet shows how to create a stream of line lengths from an input stream of text lines.

```
val lines = ssc.socketTextStream("localhost", 9999)
val lengths = lines map {line => line.length}
```

flatMap

The flatMap method returns a new DStream created by applying a user-provided function that returns 0 or more elements for each element in the source DStream.

The following code snippet shows how to create a stream of words from an input stream of lines.

```
val lines = ssc.socketTextStream("localhost", 9999)
val words = lines flatMap {line => line.split(" ")}
```

filter

The filter method returns a new DStream created by selecting only those element in the source DStream for which the user-provided input function returns true.

The following code snippet removes empty lines from an input stream of lines.

```
val lines = ssc.socketTextStream("localhost", 9999)
val nonBlankLines = lines filter {line => line.length > 0}
```

repartition

The `repartition` method returns a new DStream in which each RDD has the specified number of partitions. It allows you to distribute input data stream across a number of machines for processing. It is used to change the level of processing parallelism. More partitions increase parallelism, while fewer partitions reduce parallelism.

```
val inputStream = ssc.socketTextStream("localhost", 9999)
inputStream.repartition(10)
```

union

The `union` method returns a new DStream that contains the union of the elements in the source `DStream` and the DStream provided as input to this method.

```
val stream1 = ...
val stream2 = ...
val combinedStream = stream1.union(stream2)
```

Aggregation Transformation

The transformations described in this section perform aggregation on the underlying RDDs of a `DStream`.

count

The `count` method returns a DStream of single-element RDDs. Each RDD in the returned DStream has the count of the elements in the corresponding RDD in the source DStream.

```
val inputStream = ssc.socketTextStream("localhost", 9999)
val countsPerRdd = inputStream.count()
```

reduce

The `reduce` method returns a DStream of single-element RDDs by reducing the elements in each RDD in the source DStream. It takes a user provided reduce function as an argument.

```
val lines = ssc.socketTextStream("localhost", 9999)
val words = lines flatMap {line => line.split(" ")}
val longestWords = words reduce { (w1, w2) => if(w1.length > w2.length) w1 else w2}
```

countByValue

The `countByValue` method returns a DStream of key-value pairs, where a key in a pair is a distinct element within a batch interval and the value is its count. Thus, each RDD in the returned DStream contains the count of each distinct element in the corresponding RDD in the source DStream.

```
val lines = ssc.socketTextStream("localhost", 9999)
val words = lines flatMap {line => line.split(" ")}
val wordCounts = words.countByValue()
```

Transformations Available Only on DStream of key-value Pairs

DStreams of key-value pairs support a few other transformations in addition to the transformations available on all types of DStreams. The commonly used transformations on DStream of key-value pairs are briefly described next.

cogroup

The cogroup method returns a DStream of *(K, Seq[V], Seq[W])* when called on a DStream of *(K, Seq[V]) and (K, Seq[W])* pairs. It applies a cogroup operation between RDDs of the DStream passed as argument and those in the source DStream.

```
val lines1 = ssc.socketTextStream("localhost", 9999)
val words1 = lines1 flatMap {line => line.split(" ")}
val wordLenPairs1 = words1 map {w => (w.length, w)}
val wordsByLen1 = wordLenPairs1.groupByKey

val lines2 = ssc.socketTextStream("localhost", 9998)
val words2 = lines2 flatMap {line => line.split(" ")}
val wordLenPairs2 = words2 map {w => (w.length, w)}
val wordsByLen2 = wordLenPairs2.groupByKey

val wordsGroupedByLen = wordsByLen1.cogroup(wordsByLen2)
```

This example used the cogroup method to find the words with the same length from two DStreams.

join

The join method takes a DStream of key-value pairs as argument and returns a DStream, which is an inner join of the source DStream and the DStream provided as input. It returns a DStream of (K, (V, W)) when called on DStreams of (K, V) and (K, W) pairs.

```
val lines1 = ssc.socketTextStream("localhost", 9999)
val words1 = lines1 flatMap {line => line.split(" ")}
val wordLenPairs1 = words1 map {w => (w.length, w)}

val lines2 = ssc.socketTextStream("localhost", 9998)
val words2 = lines2 flatMap {line => line.split(" ")}
val wordLenPairs2 = words2 map {w => (w.length, w)}

val wordsSameLength = wordLenPairs1.join(wordLenPairs2)
```

This example creates two DStreams of lines of text. It then splits them into DStreams of words. Next, it creates DStreams of key-value pairs, where a key is the length of a word and value is the word itself. Finally, it joins those two DStreams.

Left outer, right outer, and full outer join operations are also available. If a DStream of key value pairs of type (K, V) is joined with another DStream of pairs of type (K, W), full outer join returns a DStream of (K, (Option[V], Option[W])), left outer join returns a DStream of (K, (V, Option[W])), and righter outer join returns a DStream of (K, (Option[V], W)).

```
val leftOuterJoinDS = wordLenPairs1.leftOuterJoin(wordLenPairs2)
val rightOuterJoinDS = wordLenPairs1.rightOuterJoin(wordLenPairs2)
val fullOuterJoinDS = wordLenPairs1.fullOuterJoin(wordLenPairs2)
```

groupByKey

The groupByKey method groups elements within each RDD of a DStream by their keys. It returns a new DStream by applying groupByKey to each RDD in the source DStream.

```
val lines = ssc.socketTextStream("localhost", 9999)
val words = lines flatMap {line => line.split(" ")}
val wordLenPairs = words map {w => (w.length, w)}
val wordsByLen = wordLenPairs.groupByKey
```

reduceByKey

The reduceByKey method returns a new DStream of key-value pairs, where the value for each key is obtained by applying a user-provided reduce function on all the values for that key within an RDD in the source DStream.

The following example counts the number of times a word occurs within each DStream micro-batch.

```
val lines = ssc.socketTextStream("localhost", 9999)
val words = lines flatMap {line => line.split(" ")}
val wordPairs = words map { word =>  (word, 1)}
val wordCounts = wordPairs.reduceByKey(_ + _)
```

Special Transformations

The transformations discussed so far allow you to specify operations on the elements in a DStream. Under the hood, DStream converts them to RDD operations. The next two transformations deviate from this model.

transform

The transform method returns a DStream by applying an *RDD => RDD* function to each RDD in the source DStream. It takes as argument a function that takes an RDD as argument and returns an RDD. Thus, it gives you direct access to the underlying RDDs of a DStream.

This method allows you to use methods provided by the RDD API, but which do not have equivalent operations in the DStream API. For example, sortBy is a transformation available in the RDD API, but not in the DStream API. If you want to sort the elements within each RDD of a DStream, you can use the transform method as shown in the following example.

```
val lines = ssc.socketTextStream("localhost", 9999)
val words = lines.flatMap{line => line.split(" ")}
val sorted = words.transform{rdd => rdd.sortBy((w)=> w)}
```

The transform method is also useful for applying machine and graph computation algorithms to data streams. The machine learning and graph processing libraries provide classes and methods that operate at the RDD level. Within the transform method, you can use the API provided by these libraries.

updateStateByKey

The updateStateByKey method allows you to create and update states for each key in a DStream of key-value pairs. You can use this method to maintain any information about each distinct key in a DStream.

For example, you can use the updateStateByKey method to keep a running count of each distinct word in a DStream, as shown in the following example.

```
// Set the context to periodically checkpoint the DStream operations for driver fault-tolerance
ssc.checkpoint("checkpoint")

val lines = ssc.socketTextStream("localhost", 9999)
val words = lines.flatMap{line => line.split(" ")}
val wordPairs = words.map{word => (word, 1)}

// create a function of type (xs: Seq[Int], prevState: Option[Int]) => Option[Int]
val updateState = (xs: Seq[Int], prevState: Option[Int]) => {
  prevState match {
    case Some(prevCount) => Some(prevCount + xs.sum)
    case None => Some(xs.sum)
  }
}

val runningCount = wordPairs.updateStateByKey(updateState)
```

The Spark Streaming library provides multiple overloaded variants of the updateStateByKey method. The simplest version of the updateStateByKey method takes a function of type *(Seq[V], Option[S]) => Option[S]* as an argument. This user-provided function takes two arguments. The first argument is a sequence of new values for a key in a DStream RDD and the second argument is previous state of the key wrapped in the Option data type. The user-provided function updates the state of a key using the new values and previous state of a key, and returns the new state wrapped in the *Option* data type. If the update function returns *None* for a key, Spark Streaming stops maintaining state for that key.

The updateStateByKey method returns a DStream of key-value pairs, where the value in a pair is the current state of the key in that pair.

Output Operations

Output operations are DStream methods that can be used to send DStream data to an output destination. An output destination can be a file, database, or another application. Output operations are executed sequentially in the order in which they are called by an application.

Saving to a File System

The commonly used DStream output operations for saving a DStream to a file system are briefly described next.

saveAsTextFiles

The saveAsTextFiles method saves a DStream to files. It generates a directory for each DStream RDD. In each directory, it generates a file for each RDD partition. Thus, the saveAsTextFiles method creates multiple directories, each containing one or more files. The directory name for each DStream RDD is generated using the current timestamp and a user-provided prefix and optional suffix.

```
val lines = ssc.socketTextStream("localhost", 9999)
val words = lines flatMap {line => line.split(" ")}
val wordPairs = words map { word =>  (word, 1)}
val wordCounts = wordPairs.reduceByKey(_ + _)
wordCounts.saveAsTextFiles("word-counts")
```

saveAsObjectFiles

The saveAsObjectFiles method saves DStream elements as serialized objects in binary SequenceFiles. Similar to the saveAsTextFile method, it stores the data for each DStream RDD in a separate directory and creates a file for each RDD partition. The directory name for each DStream RDD is generated using the current timestamp and a user-provided prefix and optional suffix.

```
val lines = ssc.socketTextStream("localhost", 9999)
val words = lines flatMap {line => line.split(" ")}
val longWords = words filter { word =>  word.length > 3}
longWords.saveAsObjectFiles("long-words")
```

saveAsHadoopFiles

The saveAsHadoopFiles method is available on DStreams of key-value pairs. It saves each RDD in the source DStream as a Hadoop file.

saveAsNewAPIHadoopFiles

Similar to the saveAsHadoopFiles method, the saveAsNewAPIHadoopFiles method saves each RDD in a DStream of key-value pairs as a Hadoop file.

Displaying on Console

The DStream class provides the print method for displaying a DStream on the console of the machine where the driver program is running.

print

The print method, as the name implies, prints the elements in each RDD in the source DStream on the machine running the driver program. By default, it shows the first ten elements in each RDD. An overloaded version of this method allows you to specify the number of elements to print.

```
val ssc = new StreamingContext(conf, Seconds(interval))
val lines = ssc.socketTextStream("localhost", 9999)
val words = lines flatMap {line => line.split(" ")}
val longWords = words filter { word =>  word.length > 3}
longWords.print(5)
```

Saving into a Database

The foreachRDD method in the DStream class can be used to save the results obtained from processing a DStream into a database.

foreachRDD

The foreachRDD method is similar to the transform method discussed earlier. It gives you access to the RDDs in a DStream. The key difference between transform and foreachRDD is that transform returns a new DStream, whereas foreachRDD does not return anything.

The foreachRDD method is a higher-order method that takes as argument a function of type *RDD => Unit.* It applies this function to each RDD in the source DStream. All RDD operations are available to this function. It is important to note that the foreachRDD method is executed on the driver node; however, the RDD transformations and actions called within foreachRDD are executed on the worker nodes.

Two things have to be kept in mind when saving a DStream into a database. First, creating a database connection is an expensive operation. It is recommended not to open and close database connections frequently. Ideally, you should re-use a database connection for storing as many elements as possible to amortize the cost of creating a connection. Second, a database connection generally cannot be serialized and sent from master to worker nodes. Since DStreams are processed on worker nodes, database connections should be created on worker nodes.

The RDD foreachPartition action can be used for storing multiple DStream elements using the same database connection. Since the foreachRDD DStream method gives you access to all RDD operations, you can call the foreachPartition RDD method within foreachRDD. Within foreachPartition, you can open a database connection and use that connection to store all elements in the source RDD partition. You can further optimize by using a connection pool library instead of opening and closing a physical connection directly.

The following code snippet implements the approach described earlier for saving a DStream to a database. It assumes that the application is using a connection pool library such as HikariCP or BoneCP. The connection pool library is wrapped in a lazily initialized singleton object named ConnectionPool, which manages a pool of database connections.

```
resultDStream.foreachRDD { rdd =>
  rdd.foreachPartition { iterator =>
    val dbConnection = ConnectionPool.getConnection()
    val statement = dbConnection.createStatement()
    iterator.foreach {element =>
                val result = statement.executeUpdate("...")
                // check the result
                ...
            }
```

```
    statement.close()
    // return connection to the pool
    dbConnection.close()
  }
}
```

Another optimization that you can do is batch the database writes. So instead of sending one database write per element, you can batch all the inserts for an RDD partition and send just one batch update to the database per RDD partition.

The `foreachRDD` method comes handy for not only saving a DStream to a database, but it is also useful for displaying the elements in a DStream in a custom format on the driver node.

Window Operation

A window operation is a DStream operation that is applied over a sliding window of data in a stream. Successive windows have one or more overlapping RDDs (see Figure 6-5). A window operation is a stateful DStream operation that combines data across multiple batches.

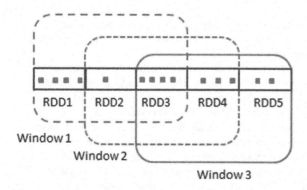

Figure 6-5. *DStream windows*

A window operation requires two parameters: window length and sliding interval (see Figure 6-6). The window length parameter specifies the time duration over which a window operation is applied. The sliding interval specifies the time interval at which a window operation is performed. It is the time interval at which new RDDs are generated by a window operation.

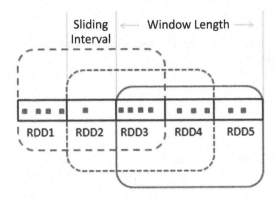

Figure 6-6. *DStream windows*

Both the window length and sliding interval parameters must be a multiple of a DStream's batch interval. The commonly used DStream methods for window operations are briefly described next.

window

The window method returns a DStream of sliding RDDs. It takes two arguments, window duration and sliding interval. Each RDD in the returned DStream includes elements from the source DStream for the specified duration and a new RDD is generated at the specified time interval. Successive RDDs in the returned DStream have overlapping data elements.

```
val lines = ssc.socketTextStream("localhost", 9999)
val words = lines flatMap {line => line.split(" ")}
val windowLen = 30
val slidingInterval = 10
val window = words.window(Seconds(windowLen), Seconds(slidingInterval))
val longestWord = window reduce { (word1, word2) =>
                        if (word1.length > word2.length) word1 else word2 }
longestWord.print()
```

countByWindow

The countByWindow method returns a DStream of single-element RDDs. The single element in each returned DStream RDD is the count of the elements in a sliding window of a specified duration. It takes two arguments, window duration, and sliding interval.

```
ssc.checkpoint("checkpoint")
val lines = ssc.socketTextStream("localhost", 9999)
val words = lines flatMap {line => line.split(" ")}
val windowLen = 30
val slidingInterval = 10
val countByWindow = words.countByWindow(Seconds(windowLen), Seconds(slidingInterval))
countByWindow.print()
```

countByValueAndWindow

The countByValueAndWindow method returns a DStream containing the counts of each distinct element within a sliding window that slides at the specified time interval.

```
ssc.checkpoint("checkpoint")
val lines = ssc.socketTextStream("localhost", 9999)
val words = lines flatMap {line => line.split(" ")}
val windowLen = 30
val slidingInterval = 10
val countByValueAndWindow = words.countByValueAndWindow(Seconds(windowLen),
Seconds(slidingInterval))
countByValueAndWindow.print()
```

reduceByWindow

The reduceByWindow method returns a DStream of single-element RDDs. Each RDD in the returned DStream is generated by applying a user-provided reduce function over the DStream elements in a sliding window. The reduceByWindow method takes three arguments: reduce function, window duration, and sliding interval.

   The user-provided reduce function must be of type *(T, T) => T*. It takes two arguments of type T and returns a single value of type T. This function is applied to all the elements within a window to generate a single value. It can be used to aggregate elements within each sliding window.

```
ssc.checkpoint("checkpoint")
val lines = ssc.socketTextStream("localhost", 9999)
val words = lines flatMap {line => line.split(" ")}
val numbers = words map {x => x.toInt}
val windowLen = 30
val slidingInterval = 10
val sumLast30Seconds = numbers.reduceByWindow({(n1, n2) => n1+n2},
                               Seconds(windowLen), Seconds(slidingInterval))
sumLast30Seconds.print()
```

reduceByKeyAndWindow

The reduceByKeyAndWindow operation is available only for DStreams of key-value pairs. It is similar to reduceByWindow, except that it does the same thing for a DStream of key-value pairs. It applies a user-provided reduce function to key-value pairs in a sliding DStream window to generate single key-value pair for each distinct key within a window.

```
ssc.checkpoint("checkpoint")
val lines = ssc.socketTextStream("localhost", 9999)
val words = lines flatMap {line => line.split(" ")}
val wordPairs = words map {word => (word, 1)}
val windowLen = 30
val slidingInterval = 10
val wordCountLast30Seconds = wordPairs.reduceByKeyAndWindow((count1: Int, count2: Int) =>
                             count1 + count2, Seconds(windowLen), Seconds(slidingInterval))
wordCountLast30Seconds.print()
```

In a windowing operation, each new window overlaps with previous window. It adds some elements to and removes some from the previous window. For example, if the window duration is 60 seconds and sliding interval is 10 seconds, each new window removes 10 seconds of data from previous window and adds 10 seconds of new data. Successive windows share 40 seconds of data. Performing complete aggregation over 60 seconds for every window is inefficient. A more efficient approach is to add the aggregate for the 10 seconds of new data to the previous window's result and remove the aggregate for the 10 seconds of data that is no longer in the new window.

Spark Streaming provides an efficient variant of the reduceByKeyAndWindow operation, which incrementally updates a sliding window by using the reduced value of the predecessor window. It requires an additional inverse reduce function as an argument. It reduces the new values that enter a windows and uses the inverse reduce function to remove the values that left the window.

```
ssc.checkpoint("checkpoint")
val lines = ssc.socketTextStream("localhost", 9999)
val words = lines flatMap {line => line.split(" ")}
val wordPairs = words map {word => (word, 1)}
val windowLen = 30
val slidingInterval = 10
def add(x: Int, y: Int): Int = x + y
def subtract(x: Int, y: Int): Int = x -y
val wordCountLast30Seconds = wordPairs.reduceByKeyAndWindow(add, subtract,
                            Seconds(windowLen), Seconds(slidingInterval))
wordCountLast30Seconds.print()
```

A Complete Spark Streaming Application

At this point, you have learned the key classes and methods provided by the Spark Streaming API. You should be able to build a distributed stream data processing application using the material discussed earlier.

In this section, let's develop a complete Spark Streaming application so that you can see how the classes and methods discussed in earlier sections come together in an application. You will create an application that shows trending Twitter hashtags.

The symbol # is called a hashtag. It is used to mark a topic in a tweet. People add the hashtag symbol before a word to categorize a tweet. A tweet may contain zero or more hashtagged words.

Twitter provides access to its global stream of tweets through a streaming API. You can learn about this API on Twitter's web site at https://dev.twitter.com/streaming/overview.

To get access to the tweets through Twitter's streaming API, you need to create a Twitter account and register your application. An application needs four pieces of authentication information to connect to Twitter's streaming API: consumer key, consumer secret, access token, and access token secret. You can obtain these from Twitter. If you have a Twitter account, sign-in or create a new account. After signing-in, register your application at https://apps.twitter.com to get all the authentication credentials.

Let's create a Spark Streaming application that tracks hashtagged words and shows the ones that are trending or gaining popularity. Application source code is shown next, followed by code explanation.

```
import org.apache.spark._
import org.apache.spark.streaming._
import org.apache.spark.streaming.twitter._
import twitter4j.Status
```

```scala
object TrendingHashTags {
  def main(args: Array[String]): Unit = {
    if (args.length < 8) {
      System.err.println("Usage: TrendingHashTags <consumer key> <consumer secret> " +
                         "<access token> <access token secret> " +
                         "<language> <batch interval> <min-threshold> <show-count> " +
                         "[<filters>]")
      System.exit(1)
    }

    val Array(consumerKey, consumerSecret, accessToken, accessTokenSecret,
                      lang, batchInterval, minThreshold, showCount ) = args.take(8)
    val filters = args.takeRight(args.length - 8)

    System.setProperty("twitter4j.oauth.consumerKey", consumerKey)
    System.setProperty("twitter4j.oauth.consumerSecret", consumerSecret)
    System.setProperty("twitter4j.oauth.accessToken", accessToken)
    System.setProperty("twitter4j.oauth.accessTokenSecret", accessTokenSecret)

    val conf = new SparkConf().setAppName("TrendingHashTags")
    val ssc = new StreamingContext(conf, Seconds(batchInterval.toInt))
    ssc.checkpoint("checkpoint")
    val tweets = TwitterUtils.createStream(ssc, None, filters)
    val tweetsFilteredByLang = tweets.filter{tweet => tweet.getLang() == lang}
    val statuses = tweetsFilteredByLang.map{ tweet => tweet.getText()}
    val words = statuses.flatMap{status => status.split("""\s+""")}
    val hashTags = words.filter{word => word.startsWith("#")}
    val hashTagPairs = hashTags.map{hashtag => (hashtag, 1)}
    val tagsWithCounts = hashTagPairs.updateStateByKey(
                        (counts: Seq[Int], prevCount: Option[Int]) =>
                            prevCount.map{c => c + counts.sum}.orElse{Some(counts.sum)}
                    )
    val topHashTags = tagsWithCounts.filter{ case(t, c) =>
                              c > minThreshold.toInt
                          }
    val sortedTopHashTags = topHashTags.transform{ rdd =>
                                rdd.sortBy({case(w, c) => c}, false)
                            }
    sortedTopHashTags.print(showCount.toInt)
    ssc.start()
    ssc.awaitTermination()
  }
}
```

Let's walk through the code.

```scala
import org.apache.spark._
import org.apache.spark.streaming._
```

These two import statements are required to import the classes and functions defined in the Spark Streaming library.

```
import org.apache.spark.streaming.twitter._
```

Twitter is an advanced source of data streams; therefore, an explicit import is required to use the classes defined in the Twitter utility library provided by Spark Streaming.

```
import twitter4j.Status
```

The Twitter-related utility classes provided by Spark Streaming use the open source Twitter4J library. Twitter4J makes it easy to integrate with the Twitter API. Since you will be using the Status class from the Twitter4J library, let's import it here.

```
object TrendingHashTags {
  def main(args: Array[String]): Unit = {
    if (args.length < 8) {
      System.err.println("Usage: TrendingHashTags <consumer key> <consumer secret> " +
                         "<access token> <access token secret> " +
                         "<language> <batch interval> <min threshold> <show-count> " +
                         "[<filters>]")
      System.exit(1)
    }

    ...
    ...
  }
}
```

This is a basic outline of a Scala application. You create a singleton object and define a main method that takes an array of Strings as an argument. This array contains the command-line arguments passed to the application.

The application checks whether the minimum number of command-line arguments has been provided; otherwise it prints a message that shows the required arguments and quits. The application requires minimum eight arguments. The first four arguments are the Twitter credentials that the application needs to connect with Twitter's streaming API. The fifth argument will be used to filter tweets by language. The sixth argument is the batch-interval in seconds for creating DStream RDDs. Each DStream RDD will include tweets for that duration. The seventh argument is a threshold for filtering hashtags by count. The eighth argument controls how many hashtags will be displayed on the console.

The remaining optional arguments will be passed to Twitter's streaming API to limit received tweets to only those matching the specified keywords. The text of a tweet and a few tweet metadata fields are considered for matches. If the keywords are enclosed in double quotes then only tweets containing all the keywords will be sent to the application; otherwise, a tweet containing any keyword from the list will be sent. For example, "android ios iphone" will look for the presence of all three words in a tweet. These words need not be next to each other in a tweet.

```
val Array(consumerKey, consumerSecret, accessToken, accessTokenSecret,
                       lang, batchInterval, minThreshold, showCount ) = args.take(8)
```

The preceding code snippet extracts the mandatory arguments to local variables, which are used in the rest of the program. It is a concise way to extract multiple command-line arguments in a single expression.

```
val filters = args.takeRight(args.length - 8)
```

The remaining optional arguments are extracted as an array of filter keywords.

```
System.setProperty("twitter4j.oauth.consumerKey", consumerKey)
System.setProperty("twitter4j.oauth.consumerSecret", consumerSecret)
System.setProperty("twitter4j.oauth.accessToken", accessToken)
System.setProperty("twitter4j.oauth.accessTokenSecret", accessTokenSecret)
```

These lines pass the authentication credentials to the Twitter4J library through system properties. You set the system properties, which will be later read by the Twitter4J library.

```
val conf = new SparkConf().setAppName("TrendingHashTags")
```

Create an instance of the SparkConf class and set the name for the application. Spark monitoring UI will show the name that you specify here.

```
val ssc = new StreamingContext(conf, Seconds(batchInterval.toInt))
```

Create an instance of the StreamingContext class. This is where you specify the batching interval or the duration for each RDD in a DStream created using this instance of the StreamingContext class. A data stream will be split into batches at the time interval passed as the second parameter to the StreamingContext constructor.

```
ssc.checkpoint("checkpoint")
```

To track trending hashtags, stateful transformations will be used. Therefore, set the context to periodically checkpoint DStream operations at the specified location. In a production application, this path should point to a reliable file system such as HDFS.

```
val tweets = TwitterUtils.createStream(ssc, None, filters)
```

Use the factory method createStream in the TwitterUtils object to create a DStream that represents a stream of tweets. The filters variable restricts the tweets to only those that match the keywords in the filters array.

```
val tweetsFilteredByLang = tweets.filter{tweet => tweet.getLang() == lang}
```

Filter the tweets to only those that match the language specified in the command-line argument. Provide a function literal to the filter transformation method, and in this function literal, use the Twitter4J API to check the language of a tweet. The getLang method in the Twitter4J's Status class returns the language of the status text, if available. Note that this method is not guaranteed to return the correct language.

```
val statuses = tweetsFilteredByLang.map{ tweet => tweet.getText()}
```

Use the map transformation to create a DStream of status texts.

```
val words = statuses.flatMap{status => status.split("""\s+""")}
```

Split the status text into words to create a DStream of words from a DStream of status texts.

```
val hashTags = words.filter{word => word.startsWith("#")}
```

Since you are interested in hashtags, filter the hashtags from the DStream of words. The `filter` method will return a DStream of hashtags.

```
val hashTagPairs = hashTags.map{hashtag => (hashtag, 1)}
```

Next, create a DStream of key-value pairs from the DStream of hashtags that you created in the previous step. This step is required since you will be using a DStream transformation that is available only for DStream of key-value pairs.

```
val tagsWithCounts = hashTagPairs.updateStateByKey(
                        (counts: Seq[Int], prevCount: Option[Int]) =>
                            prevCount.map{c => c + counts.sum}.orElse{Some(counts.sum)})
```

This is the crux of the tweet processing logic. Use the `updateStateByKey` method to track the count of each distinct hashtag in a tweet stream. A function literal is passed to the `updateStateByKey` method. This function literal updates the count for each distinct hashtag by adding the number of its occurrences in the current RDD to its previous cumulative count. The first time this function is called, there will be no previous count, so it just returns the number of times a hashtag appeared in the current RDD. The returned value is wrapped in an `Option` data type. The `updateStateByKey` method returns a `DStream` of key-value pairs, where the key in a pair is a distinct hashtag and the value is its cumulative count from the time the application started processing a Twitter stream.

```
val topHashTags = tagsWithCounts.filter{ case(t, c) =>
                            c > minThreshold.toInt
                  }
```

Next, filter the `DStream` to keep only those hashtags with a count greater than the minimum threshold that was passed to the application as a command-line argument.

```
val sortedTopHashTags = topHashTags.transform{ rdd =>
                            rdd.sortBy({case(w, c) => c}, false)
                  }
```

Here you use the `DStream` `transform` method to sort the `DStream` of hashtag-count pairs by count. Note that the DStream API itself does not provide a method to sort the elements in a DStream. Therefore, you leverage the `sortBy` method provided by the RDD API. The `transform` method returns a DStream of hashtag-count pairs sorted by count.

```
sortedTopHashTags.print(showCount.toInt)
```

Next, print the specified number of top hashtags and their counts. Note that this print statement will be executed every time a new RDD is generated for the `sortedTopHashTags` `DStream`. A new RDD will be generated as per the batch interval that you specified when you created an instance of the `StreamingContext`. The application will print the top hashtags at those intervals.

```
ssc.start()
```

Finally, start the stream computation. No processing takes place until this step. The application starts receiving and processing tweets after you call the start method.

```
ssc.awaitTermination()
```

Wait for the stream computation to stop. The stream computation will stop when either the application throws an exception or if the user interrupts the application. Otherwise, the application will continue to run forever.

Summary

Spark Streaming is a library that extends Spark for processing live data streams. It runs on Spark and provides a higher-level API for processing data streams. It turns Spark into a distributed stream data processing framework.

The primary abstraction in Spark Streaming is DStream, which represents a data stream as a never-ending sequence of RDDs. Spark Streaming uses a micro-batching architecture, in which a data stream is split into micro-batches of a specified time duration.

In addition to all the operations that Spark core provides for a micro-batch, Spark Streaming also provides windowing operations on these micro-batches. It also allows an application to maintain state information for each distinct key in a DStream of key-value pairs.

CHAPTER 7

Spark SQL

Ease of use is one of the reasons Spark became popular. It provides a simpler programming model than Hadoop MapReduce for processing big data. However, the number of people who are fluent in the languages supported by the Spark core API is a lot smaller than the number of people who know the venerable SQL.

SQL is an ANSI/ISO standard language for working with data. It specifies an interface for not only storing, modifying and retrieving data, but also for analyzing data. SQL is a declarative language. It is much easier to learn and use compared to general-purpose programming languages such as Scala, Java and Python. However, it is a powerful language for working with data. As a result, SQL has become the workhorse for data analysis.

HiveQL is a SQL-like language widely used in the Hadoop world. It is one of the preferred interfaces for Hadoop MapReduce. Instead of writing Java programs that use the low-level Hadoop MapReduce API, people prefer to use HiveQL for processing data.

This chapter covers Spark SQL, which brings the simplicity and power of SQL and HiveQL to Spark. It describes Spark SQL and its API in detail. The chapter wraps up with two detailed examples.

Introducing Spark SQL

Spark SQL is a Spark library that runs on top of Spark. It provides a higher-level abstraction than the Spark core API for processing structured data. Structured data includes data stored in a database, NoSQL data store, Parquet, ORC, Avro, JSON, CSV, or any other structured format.

Spark SQL is more than just about providing SQL interface to Spark. It was designed with the broader goals of making Spark easier to use, increasing developer productivity, and making Spark applications run faster.

Spark SQL can be used as a library for developing data processing applications in Scala, Java, Python, or R. It supports multiple query languages, including SQL, HiveQL, and language integrated queries. In addition, it can be used for interactive analytics with just SQL/HiveQL. In both cases, it internally uses the Spark core API to execute queries on a Spark cluster.

Integration with Other Spark Libraries

Spark SQL seamlessly integrates with other Spark libraries such as Spark Streaming, Spark ML, and GraphX (see Figure 7-1). It can be used for not only interactive and batch processing of historical data but also live data stream processing along with Spark Streaming. Similarly, it can be used in machine learning applications with MLlib and Spark ML. For example, Spark SQL can be used for feature engineering in a machine learning application.

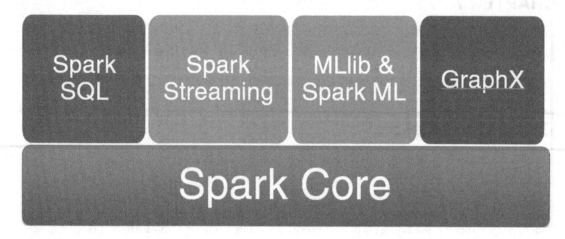

Figure 7-1. *Key Spark libraries*

Usability

Spark SQL makes Spark easier to use than the Spark core API. It provides a higher-level API and abstraction for processing structured data. For example, the Spark SQL API provides functions for selecting columns, filtering rows, aggregating columns, joining datasets, and other common data processing and analytic tasks.

One of the benefits of using Spark SQL is that it improves productivity. It enables you to process structured data with less code compared to using the Spark core API. Common data processing and analysis tasks can be implemented in Spark SQL with just a few lines of code.

Data Sources

Spark SQL supports a variety of data sources. It can be used to process data stored in a file, a NoSQL datastore, or a database. A file can be on HDFS, S3, or local file system. The file formats supported by Spark SQL include CSV, JSON, Parquet, ORC, and Avro.

Spark SQL supports a number of relational databases and NoSQL datastores. The relational databases supported by Spark SQL include PostgreSQL, MySQL, H2, Oracle, DB2, MS SQL Server, and other databases that provide JDBC connectivity. The NoSQL data stores that can be used with Spark SQL include HBase, Cassandra, Elasticsearch, Druid, and other NoSQL data stores. The list of data sources that can be used with Spark SQL keeps growing.

Data Processing Interface

Spark SQL exposes three data processing interfaces: SQL, HiveQL and language integrated queries. It translates queries written using any of these interfaces into Spark core API calls.

As previously mentioned, both SQL and HiveQL are higher-level declarative languages. In a declarative language, you just specify what you want. They are much easier to learn and use. Therefore, they are popular as the language of choice for data processing and analytics.

However, not all programmers know SQL or HiveQL. Instead of forcing these programmers to learn another language, Spark SQL supports language integrated queries in Scala, Java, Python and R. With language integrated queries, Spark SQL adds data processing capabilities to the host language; programmers can process and analyze data using the native host language syntax.

In addition, language integrated queries eliminate the impedance mismatch between SQL and the other programming languages supported by Spark. It allows a programmer to query data using SQL and process the results in Scala, Java, Python or R.

Another benefit of language integrated queries is that it reduces errors. When SQL is used for querying data, a query is specified as a string. A compiler cannot detect errors within a string. Therefore, errors within a query string are not found until an exception is thrown at runtime. Some of these errors can be eliminated by using equivalent language integrated queries.

Hive Interoperability

Spark SQL is compatible with Hive. It not only supports HiveQL, but can also access Hive metastore, SerDes, and UDFs. Therefore, if you have an existing Hive deployment, you can use Spark SQL alongside Hive. You do not need to move data or make any changes to your existing Hive metastore.

You can also replace Hive with Spark SQL to get better performance. Since Spark SQL supports HiveQL and Hive metastore, existing Hive workloads can be easily migrated to Spark SQL. HiveQL queries run much faster on Spark SQL than on Hive.

Starting with version 1.4.0, Spark SQL supports multiple versions of Hive. It can be configured to read Hive metastores created with different versions of Hive.

Note that Hive is not required to use Spark SQL. You can use Spark SQL with or without Hive. It has a built-in HiveQL parser. In addition, if you do not have an existing Hive metastore, Spark SQL creates one.

Performance

Spark SQL makes data processing applications run faster using a combination of techniques, including reduced disk I/O, in-memory columnar caching, query optimization, and code generation.

Reduced Disk I/O

Disk I/O is slow. It can be a significant contributor to query execution time. Therefore, Spark SQL reduces disk I/O wherever possible. For example, depending on the data source, it can skip non-required partitions, rows, or columns while reading data.

Partitioning

Reading an entire dataset to analyze just a chunk of the data is inefficient. For example, a query may have a filtering clause that eliminates a significant chunk of data before subsequent processing. Thus, a lot of I/O is wasted on data that is never used by an application. It can be avoided by partitioning a dataset.

Partitioning is a proven technique for improving read performance. A partitioned dataset is split into horizontal slices. Data may be partitioned by one or more columns. With partitioned datasets, Spark SQL skips the partitions that an application never uses.

Columnar Storage

A structured dataset has a tabular format. It is organized into rows and columns.

A dataset may have a large number of columns. However, an analytics application generally processes only a small percentage of the columns in a dataset. Nevertheless, if data is stored in a row-oriented storage format, all columns have to be read from disk. Reading all columns is wasteful and slows down an application. Spark SQL supports columnar storage formats such as Parquet, which allow reading of only the columns that are used in a query.

In-Memory Columnar Caching

Spark SQL allows an application to cache data in an in-memory columnar format from any data source. For example, you can use Spark SQL to cache a CSV or Avro file in memory in a columnar format.

When an application caches data in memory using Spark SQL, it caches only the required columns. In addition, Spark SQL compresses the cached columns to minimize memory usage and JVM garbage collection pressure. Use of columnar format for caching allows Spark SQL to apply efficient compression techniques such as run length encoding, delta encoding, and dictionary encoding.

Skip Rows

If a data source maintains statistical information about a dataset, Spark SQL takes advantage of it. For example, as discussed in Chapter 1, serialization formats such as Parquet and ORC store min and max values for each column in a row group or chunk of rows. Using this information, Spark SQL can skip reading chunk of rows.

Predicate Pushdown

Spark SQL also reduces disk I/O by using predicate pushdowns if a data source supports it. For example, if you read data from a relational database using Spark SQL and then apply some filtering operation to it, Spark SQL will push the filtering operation to the database. Instead of reading an entire table and then executing a filtering operation, Spark SQL will ask the database to natively execute the filtering operation. Since databases generally index data, native filtering is much faster than filtering at the application layer.

Query Optimization

Similar to database systems, Spark SQL optimizes a query before executing it. It generates an optimized physical query plan when a query is given to it for execution. It comes with a query optimizer called Catalyst, which supports both rule and cost based optimizations. It can even optimize across functions.

Spark SQL optimizes both SQL/HiveQL and language integrated queries submitted through its DataFrame API. They share the same query optimizer and execution pipeline. Thus, from a performance perspective it does not matter whether you use SQL, HiveQL or DataFrame API; they go through the same optimization steps. The DataFrame API is covered later in this chapter.

Catalyst splits query execution into four phases: analysis, logical optimization, physical planning, and code generation:

- The *analysis phase* starts with an unresolved logical plan and outputs a logical plan. An unresolved logical plan contains unresolved attributes. An unresolved attribute, for example, could be a column whose data type or source table is not yet known. Spark SQL uses rules and a catalog to resolve unbound attributes in a query expression. The Spark SQL catalog object tracks the columns and tables in all data sources.

- In the *logical optimization phase*, Spark SQL applies rule-based optimizations to the logical plan generated by the analysis phase. Rule-based optimizations include constant folding, predicate pushdown, projection pruning, null propagation, Boolean expression simplification, and other optimizations.

- The next phase is the *physical planning phase*. In this phase, Spark SQL selects an optimal physical plan using a cost-model. It takes as input the optimized logical plan generated by the logical optimization phase. Using rules, it then generates one or more physical plans that can be executed by the Spark execution engine. Next, it computes their costs and selects an optimal plan for execution. In addition, it performs rule-based physical optimizations, such as pipelining projections or filters into one Spark operation. It also pushes operations from the logical plan into data sources that support predicate or projection pushdown. Spark SQL generates optimized physical plans even for inefficient logical plans.

- The last phase is the *code generation phase*, where Spark SQL compiles parts of a query directly to Java bytecode. It uses a special Scala language feature to transform a tree representing a SQL expression to a Scala AST (Abstract Syntax Tree), which is fed to the Scala compiler at runtime to generate bytecode. Thus, it avoids using the Scala parser at runtime. This speeds up query execution. The generated code generally performs as fast as or faster than hand-tuned Scala or Java program.

One of the benefits of code generation is that you can write a Spark SQL application in any supported language without worrying about performance. For example, Python applications generally run slower than Java or Scala applications. However, a Spark SQL application written in Python will process data as fast as one written in Scala.

Applications

With its easy-to-use unified interface, high performance execution engine, and support for a variety of data sources, Spark SQL can be used for a variety of data processing and analytics tasks. Interactive data analytics is just one application of Spark SQL. This section discusses some of the other common applications of Spark SQL.

ETL (Extract Transform Load)

ETL is the process of reading data from one or more sources, applying some transformation on the data, and writing it to another data source. Conceptually, it consists of three steps: extract, transform and load. These steps need not be sequential; an application does not have to extract all the data before moving it to the transform step. Once it has extracted a portion of the data, it may run the three steps in parallel.

The extract step involves reading data from one or more operational systems. The data source could be a database, API, or a file. The source database can be a relational database or a NoSQL data store. A file can be in CSV, JSON, XML, Parquet, ORC, Avro, Protocol Buffers, or some other format.

The transform step involves cleaning and modifying the source data using some rules. For example, rows with invalid data may be dropped or columns with null values may be populated with some value. Transformations may include, concatenating two columns, splitting a column into multiple columns, encoding a column, translating a column from one encoding to a different encoding, or applying some other transformation to make the data ready for the destination system.

The load step writes the data to a destination data source. The destination can be a database or file.

Generally, ETL is used for data warehousing. Data is collected from a number of different operational systems, cleansed, transformed and stored into a data warehouse. However, ETL is not unique to data warehousing. For example, it can be used to enable sharing of data between two disparate systems. It can be used to convert data from one format to another. Similarly, migrating data from a legacy system to a new system is an ETL process.

Spark SQL is a great tool for developing an ETL application. With its support for a variety of data sources and distributed high-performance compute engine, it makes it easy to develop a fast ETL application. Moving data from one data source to another data source can be accomplished with just a few lines of code.

Data Virtualization

Spark SQL provides a unified abstraction for processing data from a variety of data sources. It can be used to process or analyze data from any supported data source using SQL, HiveQL or language integrated queries written in Scala, Java, Python or R.

Therefore, Spark SQL can be used as a data virtualization system for integrating heterogeneous data sources. The same interface can be used to process data from a file, JDBC-compliant database, or NoSQL datastore. You can even perform join operations across different data sources. For example, Spark SQL allows you to join a Parquet file with a PostgreSQL table.

Distributed JDBC/ODBC SQL Query Engine

As discussed earlier, Spark SQL can be used in two ways. First, it can be used as a library. In this mode, data processing tasks can be expressed as SQL, HiveQL or language integrated queries within a Scala, Java, Python, or R application.

Second, Spark SQL can be used as a distributed SQL query engine. It comes prepackaged with a Thrift/JDBC/ODBC server. A client application can connect to this server and submit SQL/HiveQL queries using Thrift, JDBC, or ODBC interface.

Spark SQL comes bundled a command-line client called Beeline, which can be used to submit HiveQL queries; however, the Spark JDBC/ODBC server can be queried from any application that supports JDBC/ODBC. For example, you can use it with graphical SQL clients such as SQuirrel. Similarly, it can be queried from BI and data visualization tools such Tableau, Zoomdata, and Qlik.

The Thrift JDBC/ODBC server provides two benefits. First, it allows non-programmers to use Spark. They can process and analyze data using just SQL/HiveQL. Second, it makes it easy for multiple users to share a single Spark cluster.

The Spark SQL JDBC/ODBC server looks like a database; however, unlike a database, it does not have a built-in storage engine. It is just a distributed SQL query engine, which uses Spark under the hood and can be used with a variety of a data sources.

Data Warehousing

A conventional data warehouse is essentially a database that is used to store and analyze large amounts of data. It consists of three tightly integrated components: data tables, system tables, and a SQL query engine. The data tables, as the name implies, store user data. The system tables store metadata about the data in the data tables. Both the data and system tables are stored as files using the underlying operating system's file system. The SQL query engine provides a SQL interface to store and analyze the data in the data tables. All the three components are generally packaged together as a proprietary software or appliance.

Spark SQL can be used to build an open source data warehousing solution. As discussed in the previous section, it comes prepackaged with a distributed SQL query engine, which can be paired with a variety of open source storage systems such as HDFS. Thus, it supports a modular architecture that allows users to mix and match different components in the data warehousing stack. For example, data can be stored in HDFS or S3 using a columnar format such as Parquet or ORC file format. Similarly, the query engine can be Spark SQL, Hive, Impala, Presto, or Apache Drill.

A Spark SQL-based data warehousing solution is more scalable, economical, and flexible than the proprietary data warehousing solutions. Both storage and processing capacity can be easily increased by adding more nodes to a Spark cluster. All the software components are open source and run on commodity hardware, so it is economical. Finally, it is flexible since it supports both schema-on-read and schema-on-write.

Schema-on-write systems require a schema to be defined before data can be stored. Examples of such systems include traditional databases. These systems require users to create a data model before data can be stored. The benefit of schema-on-write systems is that they provide efficient storage and allow fast interactive queries. Spark SQL supports schema-on-write through columnar formats such as Parquet and ORC. A combination of HDFS, columnar file format, and Spark SQL can be used to build a high-performance data warehousing solution.

Although schema-on-write systems enable fast queries, they have a few disadvantages. First, a schema-on-write system requires data to be modeled before it can be stored. Data modeling is not a trivial task. It requires upfront planning and good understanding of data. Second, data ingestion in a schema-on-write system is slow. Third, schemas are hard to change once a large amount of data has been stored. For example, adding a new column or changing a column type in a database with terabytes of data can be a challenge. Finally, it is difficult to store unstructured, semi-structured or multi-structured data in a schema-on-write system.

A schema-on-read system addresses the preceding issues. It allows data to be stored in its raw format. A schema is applied to data when it is read. Therefore, data can start flowing into a schema-on-read system anytime. A user can store data in its native format without worrying about how it will be queried. At query time, users can apply different schema on the stored data depending on the requirements. Schema-on-read not only enables agility but also allows complex evolving data.

One disadvantage of a schema-on-read system is that queries are slower than those executed on data stored in a schema-on-write system. Generally, schema-on-read approach is used for exploratory analytics or ETL (Extract-Transform-Load) workloads.

Since Spark SQL supports both schema-on-read and schema-on-write, it can be used for exploratory analytics, ETL and high-performance analytics. Thus, a data warehousing solution built with Spark SQL provides more flexibility than a proprietary data warehouse solution.

Application Programming Interface (API)

The Spark SQL library provides an application programming interface (API) in multiple languages. At the time of writing of this book, it supports Scala, Java, Python and R. It also allows you embed SQL/HiveQL within your application. You can even go back and forth between SQL/HiveQL and the native language API.

This section covers the Spark SQL API. It discusses the key classes and methods provided by the Spark SQL library. As I describe the classes and their methods, I will use simple examples to illustrate how they can be used. A more detailed analytics example with a real-world dataset is shown later in this chapter.

Key Abstractions

The Spark SQL API consists of three key abstractions: SQLContext, HiveContext and DataFrame. A Spark SQL application processes data using these abstractions.

SQLContext

SQLContext is the main entry point into the Spark SQL library. It is a class defined in the Spark SQL library. A Spark SQL application must create an instance of the SQLContext or HiveContext class.

SQLContext is required to create instances of the other classes provided by the Spark SQL library. It is also required to execute SQL queries.

Creating an Instance of SQLContext

An instance of the SparkContext class is required to create an instance of the SQLContext class. An application can create an instance of the SQLContext class, as shown next.

```
import org.apache.spark._
import org.apache.spark.sql._

val config = new SparkConf().setAppName("My Spark SQL app")
val sc = new SparkContext(config)
val sqlContext = new SQLContext(sc)
```

I will use the variable sqlContext in other examples in this chapter. Instead of repeating the preceding code snippet in every example, let's assume that the variable sqlContext has been created at the beginning of the program, as shown earlier.

Note that you do not need to create an instance of the SQLContext class in the Spark shell. The reason is explained in the section on HiveContext. So if you are using the Spark shell, you can skip the steps related to creating instances of SparkConf, SparkContext, and SQLContext.

Executing SQL Queries Programmatically

The SQLContext class provides a method named sql, which executes a SQL query using Spark. It takes a SQL statement as an argument and returns the result as an instance of the DataFrame class.

```
val resultSet = sqlContext.sql("SELECT count(1) FROM my_table")
```

HiveContext

HiveContext is an alternative entry point into the Spark SQL library. It extends the SQLContext class for processing data stored in Hive. It also provides a HiveQL parser. A Spark SQL application must create an instance of either this class or the SQLContext class.

HiveContext provides a superset of the functionality provided by SQLContext. The parser that comes with HiveContext is more powerful than the SQLContext parser. It can execute both HiveQL and SQL queries. It can read data from Hive tables. It also allows applications to access Hive UDFs (user-defined functions).

Note that Hive is not a requirement for using HiveContext. You can use HiveContext even if you do not have Hive installed. In fact, it is recommended to always use HiveContext since it provides a more complete parser than the SQLContext class.

If you want to process Hive tables or execute HiveQL queries on any data source, you must create an instance of the HiveContext class. It is required for working with tables defined in the Hive metastore. In addition, if you want to process existing Hive tables, add your hive-site.xml file to Spark's classpath, since HiveContext reads Hive configuration from the hive-site.xml file.

If HiveContext does not find a hive-site.xml file in the classpath, it creates the metastore_db and warehouse directories in the current directory. Therefore, you could end up with multiple copies of these directories if you launch a Spark SQL application from different directories. To avoid this, it is recommended to add hive-site.xml file to Spark's conf directory if you use HiveContext.

A sample bare bone `hive-site.xml` file is shown next.

```xml
<?xml version="1.0"?>
<?xml-stylesheet type="text/xsl" href="configuration.xsl"?>
<configuration>
  <property>
    <name>hive.metastore.warehouse.dir</name>
    <value>/path/to/hive/warehouse</value>
    <description>
      Local or HDFS directory for storing tables.
    </description>
  </property>
  <property>
    <name>javax.jdo.option.ConnectionURL</name>
    <value>jdbc:derby:;databaseName=/path/to/hive/metastore_db;create=true</value>
    <description>
      JDBC connection URL.
    </description>
  </property>
</configuration>
```

Creating an Instance of HiveContext

Similar to `SQLContext`, `HiveContext` requires a `SparkContext`. An application must first create an instance of the `SparkContext` class and then use it to create an instance of the `HiveContext` class, as shown next.

```scala
import org.apache.spark._
import org.apache.spark.sql._

val config = new SparkConf().setAppName("My Spark SQL app")
val sc = new SparkContext(config)
val hiveContext = new HiveContext(sc)
```

The Spark shell automatically creates an instance of the `HiveContext` class and makes it available in a variable named `sqlContext`. Therefore, you don't need to create an instance of `SQLContext` or `HiveContext` in the Spark shell.

Executing HiveQL Queries Programmatically

The `HiveContext` class provides a method named `sql`, which executes HiveQL queries. It takes a HiveQL statement as an argument and returns the result as a `DataFrame`.

```scala
val resultSet = hiveContext.sql("SELECT count(1) FROM my_hive_table")
```

DataFrame

`DataFrame` is Spark SQL's primary data abstraction. It represents a distributed collection of rows organized into named columns. It is inspired by DataFrames in R and Python. Conceptually, it is similar to a table in a relational database.

DataFrame is a class defined in the Spark SQL library. It provides various methods for processing and analyzing structured data. For example, it provides methods for selecting columns, filtering rows, aggregating columns, joining tables, sampling data, and other common data processing tasks.

Unlike RDD, DataFrame is schema aware. An RDD is a partitioned collection of opaque elements, whereas a DataFrame knows the names and types of the columns in a dataset. As a result, the DataFrame class is able to provide a rich domain-specific-language (DSL) for data processing.

The DataFrame API is easier to understand and use than the RDD API. However, if required, a DataFrame can also be operated on as an RDD. An RDD can be easily created from a DataFrame. Thus, the complete RDD interface is also available to process data represented by a DataFrame.

A DataFrame can be registered as a temporary table, which can be queried with SQL or HiveQL. A temporary table is available only while the application that registered it is running.

Row

Row is a Spark SQL abstraction for representing a row of data. Conceptually, it is equivalent to a relational tuple or row in a table.

Spark SQL provides factory methods to create Row objects. An example is shown next.

```
import org.apache.spark.sql._

val row1 = Row("Barack Obama", "President", "United States")
val row2 = Row("David Cameron", "Prime Minister", "United Kingdom")
```

The value of a column in a Row object can be fetched using its ordinal. Examples are shown next.

```
val presidentName = row1.getString(0)
val country = row1.getString(2)
```

You will see more examples in the next few sections.

Creating DataFrames

A DataFrame can be created in two ways. First, it can be created from a data source. Spark SQL provides builder methods for creating a DataFrame from a variety of data sources. Some of the data sources have built-in support and others require external libraries. Second, a DataFrame can be created from an RDD.

Creating a DataFrame from an RDD

Spark SQL provides two methods for creating a DataFrame from an RDD: toDF and createDataFrame.

toDF

Spark SQL provides an implicit conversion method named toDF, which creates a DataFrame from an RDD of objects represented by a case class. When this technique is used, Spark SQL infers the schema of a dataset.

The toDF method is not defined in the RDD class, but it is available through an implicit conversion. To convert an RDD to a DataFrame using toDF, you need to import the implicit methods defined in the implicits object, as shown next.

```
val sqlContext = new org.apache.spark.sql.SQLContext(sc)
import sqlContext.implicits._
```

For example, consider a CSV file that contains information about employees in a company. Each row in this file contains the name, age, and gender of an employee. The following code shows how you can use the toDF method to create a DataFrame from an RDD.

```
import org.apache.spark._
import org.apache.spark.sql._

val config = new SparkConf().setAppName("My Spark SQL app")
val sc = new SparkContext(config)
val sqlContext = new SQLContext(sc)
import sqlContext.implicits._

case class Employee(name: String, age: Int, gender: String)

val rowsRDD = sc.textFile("path/to/employees.csv")
val employeesRDD = rowsRDD.map{row => row.split(",")}
                        .map{cols => Employee(cols(0), cols(1).trim.toInt, cols(2))}

val employeesDF = employeesRDD.toDF()
```

createDataFrame

Both the SQLContext and HiveContext classes provide a method named createDataFrame, which creates a DataFrame from an RDD of Rows. The createDataFrame method takes two arguments, an RDD of Rows and a schema, and returns a DataFrame.

The schema for a dataset can be specified with an instance of StructType, which is a case class. A StructType object contains a sequence of StructField objects. StructField is also defined as a case class. It is used to specify the name and data type of a column, and optionally whether a column can contain null values and its metadata.

The org.apache.spark.sql.types package defines the various data types supported by Spark SQL along with StructType and StructField. You need to import this package to use the createDataFrame method.

Let's consider the same example used in the previous section. Assume that you have a CSV file that contains information about employees in a company. Each row in this file contains the name, age, and gender of an employee. This time you will use the createDataFrame method to create a DataFrame from an RDD.

```
import org.apache.spark._
import org.apache.spark.sql._
import org.apache.spark.sql.types._

val config = new SparkConf().setAppName("My Spark SQL app")
val sc = new SparkContext(config)
val sqlContext = new SQLContext(sc)

val linesRDD = sc.textFile("path/to/employees.csv")
val rowsRDD = linesRDD.map{row => row.split(",")}
                    .map{cols => Row(cols(0), cols(1).trim.toInt, cols(2))}
```

```
val schema = StructType(List(
                StructField("name", StringType, false),
                StructField("age", IntegerType, false),
                StructField("gender", StringType, false)
             )
           )

val employeesDF = sqlContext.createDataFrame(rowsRDD,schema)
```

The key difference between the toDF and createDataFrame methods is that the former infers the schema of a dataset and the latter requires you to specify the schema. Thus, the toDF method looks easier to use; however, the createDataFrame method provides you flexibility.

The createDataFrame method does not require you to hardcode the schema of a dataset. You can create different StructType objects at runtime. For example, you could pass schema information in a configuration file. Thus, the same application can process different datasets without requiring recompilation.

Creating a DataFrame from a Data Source

Spark SQL provides a unified interface for creating a DataFrame from a variety of data sources. For example, the same API can be used to create a DataFrame from a MySQL, PostgreSQL, Oracle, or Cassandra table. Similarly, the same API can be used to create a DataFrame from a Parquet, JSON, ORC or CSV file on local file system, HDFS or S3.

Spark SQL has built-in support for some of the commonly used data sources. These include Parquet, JSON, Hive, and JDBC-compliant databases. External packages are available for other data sources.

Spark SQL provides a class named DataFrameReader, which defines the interface for reading data from a data source. It allows you to specify different options for reading data. Through its builder methods you can specify format, partitioning, and other data source specific options. Both the SQLContext and HiveContext classes provide a factory method named read, which returns an instance of the DataFrameReader class.

The following examples illustrate how to create a DataFrame from different data sources.

```
import org.apache.spark._
import org.apache.spark.sql._

val config = new SparkConf().setAppName("My Spark SQL app")
val sc = new SparkContext(config)
val sqlContext = new org.apache.spark.sql.hive.HiveContext (sc)

// create a DataFrame from parquet files
val parquetDF = sqlContext.read
                          .format("org.apache.spark.sql.parquet")
                          .load("path/to/Parquet-file-or-directory")

// create a DataFrame from JSON files
val jsonDF = sqlContext.read
                       .format("org.apache.spark.sql.json")
                       .load("path/to/JSON-file-or-directory")

// create a DataFrame from a table in a Postgres database
val jdbcDF = sqlContext.read
                 .format("org.apache.spark.sql.jdbc")
                 .options(Map(
```

```
                    "url" -> "jdbc:postgresql://host:port/database?user=<USER>&password=<PASS>",
                    "dbtable" -> "schema-name.table-name"))
                .load()

// create a DataFrame from a Hive table
val hiveDF = sqlContext.read
                    .table("hive-table-name")
```

Similar to the methods that create an RDD, the methods that create a DataFrame are lazy. For example, the load method is lazy. Data is not read when you call the load method. An action triggers a read from a data source.

In addition to the generic method shown earlier for creating a DataFrame from a variety of data sources, Spark SQL provides special methods for creating a DataFrame from the data sources for which it has built-in support for. These data sources include Parquet, ORC, JSON, Hive, and JDBC-compliant databases.

JSON

The DataFrameReader class provides a method named json for reading a JSON dataset. It takes a path as argument and returns a DataFrame. The path can be the name of either a JSON file or a directory containing multiple JSON files.

```
val jsonDF = sqlContext.read.json("path/to/JSON-file-or-directory")
```

Spark SQL can create a DataFrame from a JSON dataset stored on any Hadoop-supported storage system; the JSON files can be on a local file system, HDFS or Amazon S3.

```
val jsonHdfsDF = sqlContext.read.json("hdfs://NAME_NODE/path/to/data.json")
val jsonS3DF = sqlContext.read.json("s3a://BUCKET_NAME/FOLDER_NAME/data.json")
```

The input JSON files must have one JSON object per line. Spark SQL will fail to read a JSON file with multi-line JSON objects.

Spark SQL automatically infers the schema of a JSON dataset. It scans the entire dataset once to determine the schema. If you know the schema, you can avoid the extra scan and speed up the DataFrame creation step by specifying the schema. For example, assume you have a JSON file containing user objects, where each object has three fields: name, age and gender of a user. You can create a DataFrame from this JSON file, as shown next.

```
import org.apache.spark.sql.types._

val userSchema = StructType(List(
                    StructField("name", StringType, false),
                    StructField("age", IntegerType, false),
                    StructField("gender", StringType, false)
                    )
                )

val userDF = sqlContext.read
                    .schema(userSchema)
                    .json("path/to/user.json")
```

Parquet

The DataFrameReader class provides a method named parquet for reading Parquet files. It takes a path of one or more Parquet files as argument and returns a DataFrame.

```
val parquetDF = sqlContext.read.parquet("path/to/parquet-file-or-directory")
```

Spark SQL can create a DataFrame from Parquet files stored on any Hadoop-supported storage system, including local file system, HDFS and Amazon S3.

```
val parquetHdfsDF = sqlContext.read.parquet("hdfs://NAME_NODE/path/to/data.parquet")
val parquetS3DF = sqlContext.read.parquet("s3a://BUCKET_NAME/FOLDER_NAME/data.parquet")
```

ORC

The DataFrameReader class provides a method named orc for reading a dataset stored in ORC file format. It takes a path as argument and returns a DataFrame.

```
val orcDF = hiveContext.read.orc("path/to/orc-file-or-directory")
```

Similar to JSON and Parquet dataset, a dataset in ORC format can be read from any Hadoop-supported storage system, including local file system, HDFS and Amazon S3.

```
val orcHdfsDF = sqlContext.read.orc("hdfs://NAME_NODE/path/to/data.orc")
val orcS3DF = sqlContext.read.orc("s3a://BUCKET_NAME/FOLDER_NAME/data.orc")
```

Hive

You can create a DataFrame from a Hive table in two ways. First, you can use the table method defined in the DataFrameReader class. It takes the name of a table in a Hive metastore and returns a DataFrame.

```
val hiveDF = hiveContext.read.table("hive-table-name")
```

Generally, a table in the Spark SQL API refers to a dataset whose metadata is persisted in a Hive metastore.

Second, you can use the sql method in the HiveContext class to create a DataFrame from a Hive table.

```
val hiveDF = hiveContext.sql("SELECT col_a, col_b, col_c from hive-table")
```

An instance of the HiveContext class is required to work with Hive tables. In addition, if you have an existing Hive deployment, copy the hive-site.xml file from your Hive setup to Spark's conf directory. The hive-site.xml file stores configuration information for Hive. For example, it has a configuration property named hive.metastore.warehouse.dir, which specifies the directory where Hive stores data for managed tables. Another important configuration property is javax.jdo.option.ConnectionURL, which specifies how to connect to the Hive metastore server.

JDBC-Compliant Database

The jdbc method defined in the DataFrameReader class creates a DataFrame from any JDBC-compliant database, including MySQL, PostgresSQL, H2, Oracle, SQL Server, SAP Hana, and DB2. The DataFrameReader class provides several overloaded variants of the jdbc method. The simplest one takes three arguments: JDBC URL of a database, table name, and connection properties. The connection properties specify connection arguments such as user name and password.

An example is shown next.

```
val jdbcUrl ="jdbc:mysql://host:port/database"
val tableName = "table-name"
val connectionProperties = new java.util.Properties
connectionProperties.setProperty("user","database-user-name")
connectionProperties.setProperty("password"," database-user-password")

val jdbcDF = hiveContext.read
                    .jdbc(jdbcUrl, tableName, connectionProperties)
```

A variant of the jdbc method allows you to specify a predicate for reading a subset of the data in a table. The predicate parameter is an array of string, where each string specifies a condition in a WHERE clause.

Assume you have a database table of user profiles, where each row stores the name, age, gender, city, state and country of a user. The following example shows how to create a DataFrame representing only users in Germany.

```
val predicates = Array("country='Germany'")

val usersGermanyDF = hiveContext.read
                    .jdbc(jdbcUrl, tableName, predicates, connectionProperties)
```

Processing Data Programmatically with SQL/HiveQL

The sql method in the HiveContext class allows you to process a dataset using HiveQL statements, whereas the sql method in the SQLContext class allows you to process a dataset using SQL statements. The table referenced in a SQL/HiveQL statement must have an entry in a Hive metastore. If you do not have an existing Hive deployment, you can create a temporary table using the registerTempTable method provided by the DataFrame class. I will discuss this method in more detail in the next section. An example is shown next.

```
import org.apache.spark.sql.types._

val userSchema = StructType(List(
                    StructField("name", StringType, false),
                    StructField("age", IntegerType, false),
                    StructField("gender", StringType, false)
                    )
                )

val userDF = sqlContext.read
                    .schema(userSchema)
                    .json("path/to/user.json")
```

```
userDF.registerTempTable("user")
val cntDF = hiveContext.sql("SELECT count(1) from user")
val cntByGenderDF = hiveContext.sql(
                        "SELECT gender, count(1) as cnt FROM user GROUP BY gender ORDER BY cnt")
```

The `sql` method returns result as a DataFrame, which provides methods for displaying the returned result on a console or saving it to a data source. The next section covers the DataFrame API in detail.

Processing Data with the DataFrame API

The previous section discussed the `sql` method in the `SQLContext` and `HiveContext` classes for processing a dataset using embedded SQL or HiveQL statements. The DataFrame API provides an alternative way for processing a dataset.

The DataFrame API consists of five types of operations, which are discussed in this section. Since you will be using a DataFrame in the examples, let's create a few.

I will use the Spark shell so that you can follow along. As a first step, launch the Spark shell from a terminal.

```
$ cd SPARK_HOME
$ ./bin/spark-shell --master local[*]
```

Once inside the Spark-shell, create a few DataFrames.

```
case class Customer(cId: Long, name: String, age: Int, gender: String)
val customers = List(Customer(1, "James", 21, "M"),
                     Customer(2, "Liz", 25, "F"),
                     Customer(3, "John", 31, "M"),
                     Customer(4, "Jennifer", 45, "F"),
                     Customer(5, "Robert", 41, "M"),
                     Customer(6, "Sandra", 45, "F"))

val customerDF = sc.parallelize(customers).toDF()

case class Product(pId: Long, name: String, price: Double, cost: Double)
val products = List(Product(1, "iPhone", 600, 400),
                    Product(2, "Galaxy", 500, 400),
                    Product(3, "iPad", 400, 300),
                    Product(4, "Kindle", 200, 100),
                    Product(5, "MacBook", 1200, 900),
                    Product(6, "Dell", 500, 400))

val productDF = sc.parallelize(products).toDF()

case class Home(city: String, size: Int, lotSize: Int,
                bedrooms: Int, bathrooms: Int, price: Int)
val homes = List(Home("San Francisco", 1500, 4000, 3, 2, 1500000),
                 Home("Palo Alto", 1800, 3000, 4, 2, 1800000),
                 Home("Mountain View", 2000, 4000, 4, 2, 1500000),
                 Home("Sunnyvale", 2400, 5000, 4, 3, 1600000),
```

```
            Home("San Jose", 3000, 6000, 4, 3, 1400000),
            Home("Fremont", 3000, 7000, 4, 3, 1500000),
            Home("Pleasanton", 3300, 8000, 4, 3, 1400000),
            Home("Berkeley", 1400, 3000, 3, 3, 1100000),
            Home("Oakland", 2200, 6000, 4, 3, 1100000),
            Home("Emeryville", 2500, 5000, 4, 3, 1200000))
```

```
val homeDF = sc.parallelize(homes).toDF
```

Instead of creating a DataFrame in every example, one of the DataFrames created here will be used. In addition, for brevity sake, the result printed by the Spark shell will be shown only for some of the examples.

As mentioned previously, the Spark shell automatically creates an instance of the SparkContext class and makes it available as sc. It also creates an instance of the HiveContext class and makes it available in a variable named sqlContext. Therefore, you do not need to create an instance of SparkContext, SQLContext or HiveContext in the Spark shell.

Basic Operations

This section describes the commonly used basic operations provided by the DataFrame class.

cache

The cache method stores the source DataFrame in memory using a columnar format. It scans only the required columns and stores them in compressed in-memory columnar format. Spark SQL automatically selects a compression codec for each column based on data statistics.

```
customerDF.cache()
```

The caching functionality can be tuned using the setConf method in the SQLContext or HiveContext class. The two configuration parameters for caching are spark.sql.inMemoryColumnarStorage.compressed and spark.sql.inMemoryColumnarStorage.batchSize. By default, compression is turned on and the batch size for columnar caching is 10,000.

```
sqlContext.setConf("spark.sql.inMemoryColumnarStorage.compressed", "true")
sqlContext.setConf("spark.sql.inMemoryColumnarStorage.batchSize", "10000")
```

columns

The columns method returns the names of all the columns in the source DataFrame as an array of String.

```
val cols = customerDF.columns
```

```
cols: Array[String] = Array(cId, name, age, gender)
```

dtypes

The dtypes method returns the data types of all the columns in the source DataFrame as an array of tuples. The first element in a tuple is the name of a column and the second element is the data type of that column.

```
val columnsWithTypes = customerDF.dtypes
```

```
columnsWithTypes: Array[(String, String)] = Array((cId,LongType), (name,StringType),
(age,IntegerType), (gender,StringType))
```

explain

The explain method prints the physical plan on the console. It is useful for debugging.

```
customerDF.explain()
```

```
== Physical Plan ==
InMemoryColumnarTableScan [cId#0L,name#1,age#2,gender#3], (InMemoryRelation
[cId#0L,name#1,age#2,gender#3], true, 10000, StorageLevel(true, true, false, true, 1),
(Scan PhysicalRDD[cId#0L,name#1,age#2,gender#3]), None)
```

A variant of the explain method takes a Boolean argument and prints both logical and physical plans if the argument is true.

persist

The persist method caches the source DataFrame in memory.

```
customerDF.persist
```

Similar to the persist method in the RDD class, a variant of the persist method in the DataFrame class allows you to specify the storage level for a persisted DataFrame.

printSchema

The printSchema method prints the schema of the source DataFrame on the console in a tree format.

```
customerDF.printSchema()
```

```
root
 |-- cId: long (nullable = false)
 |-- name: string (nullable = true)
 |-- age: integer (nullable = false)
 |-- gender: string (nullable = true)
```

registerTempTable

The registerTempTable method creates a temporary table in Hive metastore. It takes a table name as an argument.

A temporary table can be queried using the sql method in SQLContext or HiveContext. It is available only during the lifespan of the application that creates it.

```
customerDF.registerTempTable("customer")
val countDF = sqlContext.sql("SELECT count(1) AS cnt FROM customer")
```

```
countDF: org.apache.spark.sql.DataFrame = [cnt: bigint]
```

Note that the sql method returns a DataFrame. Let's discuss how to view or retrieve the contents of a DataFrame in a moment.

toDF

The toDF method allows you to rename the columns in the source DataFrame. It takes the new names of the columns as arguments and returns a new DataFrame.

```
val resultDF = sqlContext.sql("SELECT count(1) from customer")
```

```
resultDF: org.apache.spark.sql.DataFrame = [_c0: bigint]
```

```
val countDF = resultDF.toDF("cnt")
```

```
countDF: org.apache.spark.sql.DataFrame = [cnt: bigint]
```

Language-Integrated Query Methods

This section describes the commonly used language-integrated query methods of the DataFrame class.

agg

The agg method performs specified aggregations on one or more columns in the source DataFrame and returns the result as a new DataFrame.

```
val aggregates = productDF.agg(max("price"), min("price"), count("name"))
```

```
aggregates: org.apache.spark.sql.DataFrame = [max(price): double, min(price): double,
count(name): bigint]
```

```
aggregates.show
```

```
+----------+----------+-----------+
|max(price)|min(price)|count(name)|
+----------+----------+-----------+
|    1200.0|     200.0|          6|
+----------+----------+-----------+
```

Let's discuss the show method in a little bit. Essentially, it displays the content of a DataFrame on the console.

apply

The apply method takes the name of a column as an argument and returns the specified column in the source DataFrame as an instance of the Column class. The Column class provides operators for manipulating a column in a DataFrame.

```
val priceColumn = productDF.apply("price")
```

```
priceColumn: org.apache.spark.sql.Column = price
```

```
val discountedPriceColumn = priceColumn * 0.5
```

```
discountedPriceColumn: org.apache.spark.sql.Column = (price * 0.5)
```

Scala provides syntactic sugar that allows you to use productDF("price") instead of productDF.apply("price"). It automatically converts productDF("price") to productDF.apply("price"). So the preceding code can be rewritten, as shown next.

```
val priceColumn = productDF("price")
val discountedPriceColumn = priceColumn * 0.5
```

An instance of the Column class is generally used as an input to some of the DataFrame methods or functions defined in the Spark SQL library. Let's revisit one of the examples discussed earlier.

```
val aggregates = productDF.agg(max("price"), min("price"), count("name"))
```

It is a concise version of the statement shown next.

```
val aggregates = productDF.agg(max(productDF("price")), min(productDF("price")),
                              count(productDF("name")))
```

The expression productDF("price") can also be written as $"price" for convenience. Thus, the following two expressions are equivalent.

```
val aggregates = productDF.agg(max($"price"), min($"price"), count($"name"))
val aggregates = productDF.agg(max(productDF("price")), min(productDF("price")),
                              count(productDF("name")))
```

If a method or function expects an instance of the Column class as an argument, you can use the $"..." notation to select a column in a DataFrame.

In summary, the following three statements are equivalent.

```scala
val aggregates = productDF.agg(max(productDF("price")), min(productDF("price")),
                               count(productDF("name")))
val aggregates = productDF.agg(max("price"), min("price"), count("name"))
val aggregates = productDF.agg(max($"price"), min($"price"), count($"name"))
```

cube

The cube method takes the names of one or more columns as arguments and returns a cube for multi-dimensional analysis. It is useful for generating cross-tabular reports.

Assume you have a dataset that tracks sales along three dimensions: time, product and country. The cube method allows you to generate aggregates for all the possible combinations of the dimensions that you are interested in.

```scala
case class SalesSummary(date: String, product: String, country: String, revenue: Double)
val sales = List(SalesSummary("01/01/2015", "iPhone", "USA", 40000),
                SalesSummary("01/02/2015", "iPhone", "USA", 30000),
                SalesSummary("01/01/2015", "iPhone", "China", 10000),
                SalesSummary("01/02/2015", "iPhone", "China", 5000),
                SalesSummary("01/01/2015", "S6", "USA", 20000),
                SalesSummary("01/02/2015", "S6", "USA", 10000),
                SalesSummary("01/01/2015", "S6", "China", 9000),
                SalesSummary("01/02/2015", "S6", "China", 6000))

val salesDF = sc.parallelize(sales).toDF()

val salesCubeDF = salesDF.cube($"date", $"product", $"country").sum("revenue")
```

```
salesCubeDF: org.apache.spark.sql.DataFrame = [date: string, product: string, country:
string, sum(revenue): double]
```

```scala
salesCubeDF.withColumnRenamed("sum(revenue)", "total").show(30)
```

```
+----------+-------+-------+--------+
|      date|product|country|   total|
+----------+-------+-------+--------+
|01/01/2015|   null|    USA| 60000.0|
|01/02/2015|     S6|   null| 16000.0|
|01/01/2015| iPhone|   null| 50000.0|
|01/01/2015|     S6|  China|  9000.0|
|      null|   null|  China| 30000.0|
|01/02/2015|     S6|    USA| 10000.0|
|01/02/2015|   null|   null| 51000.0|
|01/02/2015| iPhone|  China|  5000.0|
|01/01/2015| iPhone|    USA| 40000.0|
|01/01/2015|   null|  China| 19000.0|
|01/02/2015|   null|    USA| 40000.0|
```

```
|        null|  iPhone|   China|  15000.0|
|01/02/2015|      S6|   China|   6000.0|
|01/01/2015|  iPhone|   China|  10000.0|
|01/02/2015|    null|   China|  11000.0|
|        null|  iPhone|    null|  85000.0|
|        null|  iPhone|     USA|  70000.0|
|        null|      S6|    null|  45000.0|
|        null|      S6|     USA|  30000.0|
|01/01/2015|      S6|    null|  29000.0|
|        null|    null|    null| 130000.0|
|01/02/2015|  iPhone|    null|  35000.0|
|01/01/2015|      S6|     USA|  20000.0|
|        null|    null|     USA| 100000.0|
|01/01/2015|    null|    null|  79000.0|
|        null|      S6|   China|  15000.0|
|01/02/2015|  iPhone|     USA|  30000.0|
+----------+-------+-------+--------+
```

If you wanted to find the total sales of all products in the USA, you can use the following expression.

```
salesCubeDF.filter("product IS null AND date IS null AND country='USA'").show
```

```
+----+-------+-------+------------+
|date|product|country|sum(revenue)|
+----+-------+-------+------------+
|null|   null|    USA|    100000.0|
+----+-------+-------+------------+
```

If you wanted to know the subtotal of sales by product in the USA, you can use the following expression.

```
salesCubeDF.filter("date IS null AND product IS NOT null AND country='USA'").show
```

```
+----+-------+-------+------------+
|date|product|country|sum(revenue)|
+----+-------+-------+------------+
|null| iPhone|    USA|     70000.0|
|null|     S6|    USA|     30000.0|
+----+-------+-------+------------+
```

distinct

The distinct method returns a new DataFrame containing only the unique rows in the source DataFrame.

```
val dfWithoutDuplicates = customerDF.distinct
```

explode

The explode method generates zero or more rows from a column using a user-provided function. It takes three arguments. The first argument is the input column, the second argument is the output column and the third argument is a user provided function that generates one or more values for the output column for each value in the input column.

For example, consider a dataset that has a text column containing contents of an email. Let's assume that you want to split the email content into individual words and you want a row for each word in an email.

```
case class Email(sender: String, recepient: String, subject: String, body: String)
val emails = List(Email("James", "Mary", "back", "just got back from vacation"),
                  Email("John", "Jessica", "money", "make million dollars"),
                  Email("Tim", "Kevin", "report", "send me sales report ASAP"))

val emailDF = sc.parallelize(emails).toDF()
val wordDF = emailDF.explode("body", "word") { body: String => body.split(" ")}
wordDF.show
```

```
+------+---------+-------+--------------------+--------+
|sender|recepient|subject|                body|    word|
+------+---------+-------+--------------------+--------+
| James|     Mary|   back|just got back fro...|    just|
| James|     Mary|   back|just got back fro...|     got|
| James|     Mary|   back|just got back fro...|    back|
| James|     Mary|   back|just got back fro...|    from|
| James|     Mary|   back|just got back fro...|vacation|
|  John|  Jessica|  money|make million dollars|    make|
|  John|  Jessica|  money|make million dollars| million|
|  John|  Jessica|  money|make million dollars| dollars|
|   Tim|    Kevin| report|send me sales rep...|    send|
|   Tim|    Kevin| report|send me sales rep...|      me|
|   Tim|    Kevin| report|send me sales rep...|   sales|
|   Tim|    Kevin| report|send me sales rep...|  report|
|   Tim|    Kevin| report|send me sales rep...|    ASAP|
+------+---------+-------+--------------------+--------+
```

filter

The filter method filters rows in the source DataFrame using a SQL expression provided to it as an argument. It returns a new DataFrame containing only the filtered rows. The SQL expression can be passed as a string argument.

```
val filteredDF = customerDF.filter("age > 25")
```

```
filteredDF: org.apache.spark.sql.DataFrame = [cId: bigint, name: string, age: int, gender: string]
```

filteredDF.show

```
+---+--------+---+------+
|cId|    name|age|gender|
+---+--------+---+------+
|  3|    John| 31|     M|
|  4|Jennifer| 45|     F|
|  5|  Robert| 41|     M|
|  6|  Sandra| 45|     F|
+---+--------+---+------+
```

A variant of the filter method allows a filter condition to be specified using the Column type.

```
val filteredDF = customerDF.filter($"age" > 25)
```

As mentioned earlier, the preceding code is a short-hand for the following code.

```
val filteredDF = customerDF.filter(customerDF("age") > 25)
```

groupBy

The groupBy method groups the rows in the source DataFrame using the columns provided to it as arguments. Aggregation can be performed on the grouped data returned by this method.

```
val countByGender = customerDF.groupBy("gender").count
```

```
countByGender: org.apache.spark.sql.DataFrame = [gender: string, count: bigint]
```

countByGender.show

```
+------+-----+
|gender|count|
+------+-----+
|     F|    3|
|     M|    3|
+------+-----+
```

```
val revenueByProductDF = salesDF.groupBy("product").sum("revenue")
```

```
revenueByProductDF: org.apache.spark.sql.DataFrame = [product: string, sum(revenue): double]
```

revenueByProductDF.show

```
+-------+------------+
|product|sum(revenue)|
+-------+------------+
| iPhone|     85000.0|
|     S6|     45000.0|
+-------+------------+
```

intersect

The intersect method takes a DataFrame as an argument and returns a new DataFrame containing only the rows in both the input and source DataFrame.

```
val customers2 = List(Customer(11, "Jackson", 21, "M"),
                      Customer(12, "Emma", 25, "F"),
                      Customer(13, "Olivia", 31, "F"),
                      Customer(4, "Jennifer", 45, "F"),
                      Customer(5, "Robert", 41, "M"),
                      Customer(6, "Sandra", 45, "F"))

val customer2DF = sc.parallelize(customers2).toDF()
val commonCustomersDF = customerDF.intersect(customer2DF)
commonCustomersDF.show
```

```
+---+--------+---+------+
|cId|    name|age|gender|
+---+--------+---+------+
|  6|  Sandra| 45|     F|
|  4|Jennifer| 45|     F|
|  5|  Robert| 41|     M|
+---+--------+---+------+
```

join

The join method performs a SQL join of the source DataFrame with another DataFrame. It takes three arguments, a DataFrame, a join expression and a join type.

```
case class Transaction(tId: Long, custId: Long, prodId: Long, date: String, city: String)
val transactions = List(Transaction(1, 5, 3, "01/01/2015", "San Francisco"),
                        Transaction(2, 6, 1, "01/02/2015", "San Jose"),
                        Transaction(3, 1, 6, "01/01/2015", "Boston"),
                        Transaction(4, 200, 400, "01/02/2015", "Palo Alto"),
                        Transaction(6, 100, 100, "01/02/2015", "Mountain View"))

val transactionDF = sc.parallelize(transactions).toDF()
val innerDF = transactionDF.join(customerDF, $"custId" === $"cId", "inner")
```

innerDF.show

```
+---+------+------+----------+-------------+---+------+---+------+
|tId|custId|prodId|      date|         city|cId|  name|age|gender|
+---+------+------+----------+-------------+---+------+---+------+
|  1|     5|     3|01/01/2015|San Francisco|  5|Robert| 41|     M|
|  2|     6|     1|01/02/2015|     San Jose|  6|Sandra| 45|     F|
|  3|     1|     6|01/01/2015|       Boston|  1| James| 21|     M|
+---+------+------+----------+-------------+---+------+---+------+
```

val outerDF = transactionDF.join(customerDF, $"custId" === $"cId", "outer")
outerDF.show

```
+----+------+------+----------+-------------+----+--------+----+------+
| tId|custId|prodId|      date|         city| cId|    name| age|gender|
+----+------+------+----------+-------------+----+--------+----+------+
|   6|   100|   100|01/02/2015|Mountain View|null|    null|null|  null|
|   4|   200|   400|01/02/2015|    Palo Alto|null|    null|null|  null|
|   3|     1|     6|01/01/2015|       Boston|   1|   James|  21|     M|
|null|  null|  null|      null|         null|   2|     Liz|  25|     F|
|null|  null|  null|      null|         null|   3|    John|  31|     M|
|null|  null|  null|      null|         null|   4|Jennifer|  45|     F|
|   1|     5|     3|01/01/2015|San Francisco|   5|  Robert|  41|     M|
|   2|     6|     1|01/02/2015|     San Jose|   6|  Sandra|  45|     F|
+----+------+------+----------+-------------+----+--------+----+------+
```

val leftOuterDF = transactionDF.join(customerDF, $"custId" === $"cId", "left_outer")
leftOuterDF.show

```
+---+------+------+----------+-------------+----+------+----+------+
|tId|custId|prodId|      date|         city| cId|  name| age|gender|
+---+------+------+----------+-------------+----+------+----+------+
|  1|     5|     3|01/01/2015|San Francisco|   5|Robert|  41|     M|
|  2|     6|     1|01/02/2015|     San Jose|   6|Sandra|  45|     F|
|  3|     1|     6|01/01/2015|       Boston|   1| James|  21|     M|
|  4|   200|   400|01/02/2015|    Palo Alto|null|  null|null|  null|
|  6|   100|   100|01/02/2015|Mountain View|null|  null|null|  null|
+---+------+------+----------+-------------+----+------+----+------+
```

```
val rightOuterDF = transactionDF.join(customerDF, $"custId" === $"cId", "right_outer")
rightOuterDF.show
```

```
+----+------+------+----------+-------------+---+--------+---+------+
| tId|custId|prodId|      date|         city|cId|    name|age|gender|
+----+------+------+----------+-------------+---+--------+---+------+
|   3|     1|     6|01/01/2015|       Boston|  1|   James| 21|     M|
|null|  null|  null|      null|         null|  2|    Liz| 25|     F|
|null|  null|  null|      null|         null|  3|    John| 31|     M|
|null|  null|  null|      null|         null|  4|Jennifer| 45|     F|
|   1|     5|     3|01/01/2015|San Francisco|  5|  Robert| 41|     M|
|   2|     6|     1|01/02/2015|    San Jose|  6|  Sandra| 45|     F|
+----+------+------+----------+-------------+---+--------+---+------+
```

limit

The limit method returns a DataFrame containing the specified number of rows from the source DataFrame.

```
val fiveCustomerDF = customerDF.limit(5)
fiveCustomer.show
```

```
+---+--------+---+------+
|cId|    name|age|gender|
+---+--------+---+------+
|  1|   James| 21|     M|
|  2|    Liz| 25|     F|
|  3|    John| 31|     M|
|  4|Jennifer| 45|     F|
|  5|  Robert| 41|     M|
+---+--------+---+------+
```

orderBy

The orderBy method returns a DataFrame sorted by the given columns. It takes the names of one or more columns as arguments.

```
val sortedDF = customerDF.orderBy("name")
sortedDF.show
```

```
+---+--------+---+------+
|cId|    name|age|gender|
+---+--------+---+------+
|  1|   James| 21|     M|
|  4|Jennifer| 45|     F|
|  3|    John| 31|     M|
|  2|    Liz| 25|     F|
|  5|  Robert| 41|     M|
|  6|  Sandra| 45|     F|
+---+--------+---+------+
```

By default, the orderBy method sorts in ascending order. You can explicitly specify the sorting order using a Column expression, as shown next.

```
val sortedByAgeNameDF = customerDF.sort($"age".desc, $"name".asc)
sortedByAgeNameDF.show
```

```
+---+--------+---+------+
|cId|    name|age|gender|
+---+--------+---+------+
|  4|Jennifer| 45|     F|
|  6|  Sandra| 45|     F|
|  5|  Robert| 41|     M|
|  3|    John| 31|     M|
|  2|     Liz| 25|     F|
|  1|   James| 21|     M|
+---+--------+---+------+
```

randomSplit

The randomSplit method splits the source DataFrame into multiple DataFrames. It takes an array of weights as argument and returns an array of DataFrames. It is a useful method for machine learning, where you want to split the raw dataset into training, validation and test datasets.

```
val dfArray = homeDF.randomSplit(Array(0.6, 0.2, 0.2))
dfArray(0).count
dfArray(1).count
dfArray(2).count
```

rollup

The rollup method takes the names of one or more columns as arguments and returns a multi-dimensional rollup. It is useful for subaggregation along a hierarchical dimension such as geography or time.

Assume you have a dataset that tracks annual sales by city, state and country. The rollup method can be used to calculate both grand total and subtotals by city, state, and country.

```
case class SalesByCity(year: Int, city: String, state: String,
                       country: String, revenue: Double)
val salesByCity = List(SalesByCity(2014, "Boston", "MA", "USA", 2000),
                       SalesByCity(2015, "Boston", "MA", "USA", 3000),
                       SalesByCity(2014, "Cambridge", "MA", "USA", 2000),
                       SalesByCity(2015, "Cambridge", "MA", "USA", 3000),
                       SalesByCity(2014, "Palo Alto", "CA", "USA", 4000),
                       SalesByCity(2015, "Palo Alto", "CA", "USA", 6000),
                       SalesByCity(2014, "Pune", "MH", "India", 1000),
                       SalesByCity(2015, "Pune", "MH", "India", 1000),
                       SalesByCity(2015, "Mumbai", "MH", "India", 1000),
                       SalesByCity(2014, "Mumbai", "MH", "India", 2000))
```

```
val salesByCityDF = sc.parallelize(salesByCity).toDF()
val rollup = salesByCityDF.rollup($"country", $"state", $"city").sum("revenue")
rollup.show
```

```
+-------+-----+---------+------------+
|country|state|     city|sum(revenue)|
+-------+-----+---------+------------+
|  India|   MH|   Mumbai|      3000.0|
|    USA|   MA|Cambridge|      5000.0|
|  India|   MH|     Pune|      2000.0|
|    USA|   MA|   Boston|      5000.0|
|    USA|   MA|     null|     10000.0|
|    USA| null|     null|     20000.0|
|    USA|   CA|     null|     10000.0|
|   null| null|     null|     25000.0|
|  India|   MH|     null|      5000.0|
|    USA|   CA|Palo Alto|     10000.0|
|  India| null|     null|      5000.0|
+-------+-----+---------+------------+
```

sample

The sample method returns a DataFrame containing the specified fraction of the rows in the source DataFrame. It takes two arguments. The first argument is a Boolean value indicating whether sampling should be done with replacement. The second argument specifies the fraction of the rows that should be returned.

```
val sampleDF = homeDF.sample(true, 0.10)
```

select

The select method returns a DataFrame containing only the specified columns from the source DataFrame.

```
val namesAgeDF = customerDF.select("name", "age")
namesAgeDF.show
```

```
+--------+---+
|    name|age|
+--------+---+
|   James| 21|
|     Liz| 25|
|    John| 31|
|Jennifer| 45|
|  Robert| 41|
|  Sandra| 45|
+--------+---+
```

A variant of the select method allows one or more Column expressions as arguments.

```
val newAgeDF = customerDF.select($"name", $"age" + 10)
newAgeDF.show
```

```
+--------+----------+
|    name|(age + 10)|
+--------+----------+
|   James|        31|
|     Liz|        35|
|    John|        41|
|Jennifer|        55|
|  Robert|        51|
|  Sandra|        55|
+--------+----------+
```

selectExpr

The selectExpr method accepts one or more SQL expressions as arguments and returns a DataFrame generated by executing the specified SQL expressions.

```
val newCustomerDF = customerDF.selectExpr("name", "age + 10  AS new_age",
                                 "IF(gender = 'M', true, false) AS male")
```

```
newCustomerDF.show
```

```
+--------+-------+-----+
|    name|new_age| male|
+--------+-------+-----+
|   James|     31| true|
|     Liz|     35|false|
|    John|     41| true|
|Jennifer|     55|false|
|  Robert|     51| true|
|  Sandra|     55|false|
+--------+-------+-----+
```

withColumn

The withColumn method adds a new column to or replaces an existing column in the source DataFrame and returns a new DataFrame. It takes two arguments. The first argument is the name of the new column and the second argument is an expression for generating the values of the new column.

```
val newProductDF = productDF.withColumn("profit", $"price" - $"cost")
newProductDF.show
```

```
+---+-------+------+-----+------+
|pId|   name| price| cost|profit|
+---+-------+------+-----+------+
|  1| iPhone| 600.0|400.0| 200.0|
|  2| Galaxy| 500.0|400.0| 100.0|
|  3|    iPad| 400.0|300.0| 100.0|
|  4| Kindle| 200.0|100.0| 100.0|
|  5|MacBook|1200.0|900.0| 300.0|
|  6|    Dell| 500.0|400.0| 100.0|
+---+-------+------+-----+------+
```

RDD Operations

The DataFrame class supports commonly used RDD operations such as map, flatMap, foreach, foreachPartition, mapPartition, coalesce, and repartition. These methods work similar to their namesake operations in the RDD class.

In addition, if you need access to other RDD methods that are not present in the DataFrame class, you can get an RDD from a DataFrame. This section discusses the commonly used techniques for generating an RDD from a DataFrame.

rdd

rdd is defined as a *lazy val* in the DataFrame class. It represents the source DataFrame as an RDD of Row instances.

As discussed earlier, a Row represents a relational tuple in the source DataFrame. It allows both generic access and native primitive access of fields by their ordinal.

An example is shown next.

```
val rdd = customerDF.rdd
```

```
rdd: org.apache.spark.rdd.RDD[org.apache.spark.sql.Row] = MapPartitionsRDD[405] at rdd at
<console>:27
```

```
val firstRow = rdd.first
```

```
firstRow: org.apache.spark.sql.Row = [1,James,21,M]
```

```
val name = firstRow.getString(1)
```

```
name: String = James
```

```
val age = firstRow.getInt(2)
```

```
age: Int = 21
```

Fields in a Row can also be extracted using Scala pattern matching.

```
import org.apache.spark.sql.Row
val rdd = customerDF.rdd
```

```
rdd: org.apache.spark.rdd.RDD[org.apache.spark.sql.Row] = MapPartitionsRDD[113] at rdd at
<console>:28
```

```
val nameAndAge = rdd.map {
              case Row(cId: Long, name: String, age: Int, gender: String) => (name, age)
              }
```

```
nameAndAge: org.apache.spark.rdd.RDD[(String, Int)] = MapPartitionsRDD[114] at map at
<console>:30
```

```
nameAndAge.collect
```

```
res79: Array[(String, Int)] = Array((James,21), (Liz,25), (John,31), (Jennifer,45),
(Robert,41), (Sandra,45))
```

toJSON

The toJSON method generates an RDD of JSON strings from the source DataFrame. Each element in the returned RDD is a JSON object.

```
val jsonRDD = customerDF.toJSON
```

```
jsonRDD: org.apache.spark.rdd.RDD[String] = MapPartitionsRDD[408] at toJSON at
<console>:28
```

```
jsonRDD.collect
```

```
res80: Array[String] = Array({"cId":1,"name":"James","age":21,"gender":"M"},
{"cId":2,"name":"Liz","age":25,"gender":"F"},
{"cId":3,"name":"John","age":31,"gender":"M"},
{"cId":4,"name":"Jennifer","age":45,"gender":"F"},
{"cId":5,"name":"Robert","age":41,"gender":"M"},
{"cId":6,"name":"Sandra","age":45,"gender":"F"})
```

Actions

Similar to the RDD actions, the action methods in the `DataFrame` class return results to the Driver program. This section covers the commonly used action methods in the `DataFrame` class.

collect

The `collect` method returns the data in a DataFrame as an array of Rows.

```
val result = customerDF.collect
```

```
result: Array[org.apache.spark.sql.Row] = Array([1,James,21,M], [2,Liz,25,F],
[3,John,31,M], [4,Jennifer,45,F], [5,Robert,41,M], [6,Sandra,45,F])
```

count

The `count` method returns the number of rows in the source `DataFrame`.

```
val count = customerDF.count
```

```
count: Long = 6
```

describe

The `describe` method can be used for exploratory data analysis. It returns summary statistics for numeric columns in the source `DataFrame`. The summary statistics includes min, max, count, mean, and standard deviation. It takes the names of one or more columns as arguments.

```
val summaryStatsDF = productDF.describe("price", "cost")
```

```
summaryStatsDF: org.apache.spark.sql.DataFrame = [summary: string, price: string, cost: string]
```

```
summaryStatsDF.show
```

```
+-------+------------------+------------------+
|summary|             price|              cost|
+-------+------------------+------------------+
|  count|                 6|                 6|
|   mean|  566.6666666666666|  416.6666666666667|
| stddev|309.12061651652357|240.94720491334928|
|    min|             200.0|             100.0|
|    max|            1200.0|             900.0|
+-------+------------------+------------------+
```

first

The first method returns the first row in the source DataFrame.

```
val first = customerDF.first
```

```
first: org.apache.spark.sql.Row = [1,James,21,M]
```

show

The show method displays the rows in the source DataFrame on the driver console in a tabular format. It optionally takes an integer N as an argument and displays the top N rows. If no argument is provided, it shows the top 20 rows.

```
customerDF.show(2)
```

```
+---+-----+---+------+
|cId| name|age|gender|
+---+-----+---+------+
|  1|James| 21|     M|
|  2|  Liz| 25|     F|
+---+-----+---+------+
only showing top 2 rows
```

take

The take method takes an integer N as an argument and returns the first N rows from the source DataFrame as an array of Rows.

```
val first2Rows = customerDF.take(2)
```

```
first2Rows: Array[org.apache.spark.sql.Row] = Array([1,James,21,M], [2,Liz,25,F])
```

Output Operations

An output operation saves a DataFrame to a storage system. Prior to version 1.4, DataFrame included a number of different methods for saving a DataFrame to a variety of storage systems. Starting with version 1.4, those methods were replaced by the write method.

write

The write method returns an instance of the DataFrameWriter class, which provides methods for saving the contents of a DataFrame to a data source. The next section covers the DataFrameWriter class.

Saving a DataFrame

Spark SQL provides a unified interface for saving a DataFrame to a variety of data sources. The same interface can be used to write data to relational databases, NoSQL data stores and a variety of file formats.

The DataFrameWriter class defines the interface for writing data to a data source. Through its builder methods, it allows you to specify different options for saving data. For example, you can specify format, partitioning, and handling of existing data.

The following examples show how to save a DataFrame to different storage systems.

```
// save a DataFrame in JSON format
customerDF.write
  .format("org.apache.spark.sql.json")
  .save("path/to/output-directory")

// save a DataFrame in Parquet format
homeDF.write
  .format("org.apache.spark.sql.parquet")
  .partitionBy("city")
  .save("path/to/output-directory")

// save a DataFrame in ORC file format
homeDF.write
  .format("orc")
  .partitionBy("city")
  .save("path/to/output-directory")

// save a DataFrame as a Postgres database table
df.write
  .format("org.apache.spark.sql.jdbc")
  .options(Map(
    "url" -> "jdbc:postgresql://host:port/database?user=<USER>&password=<PASS>",
    "dbtable" -> "schema-name.table-name"))
  .save()

// save a DataFrame to a Hive table
df.write.saveAsTable("hive-table-name")
```

You can save a DataFrame in Parquet, JSON, ORC, or CSV format to any Hadoop-supported storage system, including local file system, HDFS or Amazon S3.

If a data source supports partitioned layout, the DataFrameWriter class supports it through the partitionBy method. It will partition the rows by the specified column and create a separate subdirectory for each unique value in the specified column. Partitioned layout enables partition pruning during query. Future queries through Spark SQL will be able to skip large amounts of disk I/O when a partitioned column is referenced in a predicate.

Consider the following example.

```
homeDF.write
  .format("parquet")
  .partitionBy("city")
  .save("homes")
```

The preceding code partitions rows by the city column. A subdirectory will be created for each unique value of city. For example, subdirectories named `city=Berkeley`, `city=Fremont`, `city=Oakland`, and so on, will be created under the homes directory.

The following code will only read the subdirectory named `city=Berkeley` and skip all other subdirectories under the homes directory. For a large dataset that includes data for hundreds of cities, this could speed up application performance by orders of magnitude.

```
val newHomesDF = sqlContext.read.format("parquet").load("homes")
newHomesDF.registerTempTable("homes")
val homesInBerkeley = sqlContext.sql("SELECT * FROM homes WHERE city = 'Berkeley'")
```

While saving a DataFrame, if the destination path or table already exists, Spark SQL will throw an exception by default. You can change this behavior by calling the *mode* method in the `DataFrameWriter` class. It takes an argument, which specifies the behavior if destination path or table already exists. The mode method supports the following options:

> `error` (default) - throw an exception if destination path/table already exists.
>
> `append` - append to existing data if destination path/table already exists.
>
> `overwrite` - overwrite existing data if destination path/table exists.
>
> `ignore` - ignore the operation if destination path/table exists.

A few examples are shown next.

```
customerDF.write
  .format("parquet")
  .mode("overwrite")
  .save("path/to/output-directory")

customerDF.write
  .mode("append")
  .saveAsTable("hive-table-name")
```

In addition to the methods shown here for writing data to a data source, the `DataFrameWriter` class provides special methods for writing data to the data sources for which it has built-in support for. These data sources include Parquet, ORC, JSON, Hive, and JDBC-compliant databases.

JSON

The `json` method saves the contents of a DataFrame in JSON format. It takes as argument a path, which can be on a local file system, HDFS or S3.

```
customerDF.write.json("path/to/directory")
```

Parquet

The `parquet` method saves the contents of a DataFrame in Parquet format. It takes a path as argument and saves a DataFrame at the specified path.

```
customerDF.write.parquet("path/to/directory")
```

ORC

The orc method saves a DataFrame in ORC file format. Similar to the JSON and Parquet methods, it takes a path as argument.

```
customerDF.write.orc("path/to/directory")
```

Hive

The saveAsTable method saves the content of a DataFrame as a Hive table. It saves a DataFrame to a file and registers metadata as a table in Hive metastore.

```
customerDF.write.saveAsTable("hive-table-name")
```

JDBC-Compliant Database

The jdbc method saves a DataFrame to a database using the JDBC interface. It takes three arguments: JDBC URL of a database, table name, and connection properties. The connection properties specify connection arguments such as user name and password.

```
val jdbcUrl ="jdbc:mysql://host:port/database"
val tableName = "table-name"
val connectionProperties = new java.util.Properties
connectionProperties.setProperty("user","database-user-name")
connectionProperties.setProperty("password"," database-user-password")

customerDF.write.jdbc(jdbcUrl, tableName, connectionProperties)
```

Built-in Functions

Spark SQL comes with a comprehensive list of built-in functions, which are optimized for fast execution. It implements these functions with code generation techniques. The built-in functions can be used from both the DataFrame API and SQL interface.

To use Spark's built-in functions from the DataFrame API, you need to add the following import statement to your source code.

```
import org.apache.spark.sql.functions._
```

The built-in functions can be classified into the following categories: aggregate, collection, date/time, math, string, window, and miscellaneous functions.

Aggregate

The aggregate functions can be used to perform aggregations on a column. The built-in aggregate functions include approxCountDistinct, avg, count, countDistinct, first, last, max, mean, min, sum, *and* sumDistinct.

The following example illustrates how you can use a built-in function.

```
val minPrice = homeDF.select(min($"price"))
minPrice.show
```

```
+----------+
|min(price)|
+----------+
|   1100000|
+----------+
```

Collection

The collection functions operate on columns containing a collection of elements. The built-in collection functions include array_contains, explode, size, and sort_array.

Date/Time

The date/time functions make it easy to process columns containing date/time values. These functions can be further sub-classified into the following categories: conversion, extraction, arithmetic, and miscellaneous functions.

Conversion

The conversion functions convert date/time values from one format to another. For example, you can convert a timestamp string in *yyyy-MM-dd HH:mm:ss* format to a Unix epoch value using the unix_timestamp function. Conversely, the from_unixtime function converts a Unix epoch value to a string representation. Spark SQL also provides functions for converting timestamps from one time zone to another.

The built-in conversion functions include unix_timestamp, from_unixtime, to_date, quarter, day, dayofyear, weekofyear, from_utc_timestamp, and to_utc_timestamp.

Field Extraction

The field extraction functions allow you to extract year, month, day, hour, minute, and second from a Date/Time value. The built-in field extraction functions include year, quarter, month, weekofyear, dayofyear, dayofmonth, hour, minute, and second.

Date Arithmetic

The arithmetic functions allow you to perform arithmetic operation on columns containing dates. For example, you can calculate the difference between two dates, add days to a date, or subtract days from a date. The built-in date arithmetic functions include datediff, date_add, date_sub, add_months, last_day, next_day, and months_between.

Miscellaneous

In addition to the functions mentioned earlier, Spark SQL provides a few other useful date- and time-related functions, such as `current_date`, `current_timestamp`, `trunc`, `date_format`.

Math

The math functions operate on columns containing numerical values. Spark SQL comes with a long list of built-in math functions. Example include `abs`, `ceil`, `cos`, `exp`, `factorial`, `floor`, `hex`, `hypot`, `log`, `log10`, `pow`, `round`, `shiftLeft`, `sin`, `sqrt`, `tan`, and other commonly used math functions.

String

Spark SQL provides a variety of built-in functions for processing columns that contain string values. For example, you can split, trim or change case of a string. The built-in string functions include `ascii`, `base64`, `concat`, `concat_ws`, `decode`, `encode`, `format_number`, `format_string`, `get_json_object`, `initcap`, `instr`, `length`, `levenshtein`, `locate`, `lower`, `lpad`, `ltrim`, `printf`, `regexp_extract`, `regexp_replace`, `repeat`, `reverse`, `rpad`, `rtrim`, `soundex`, `space`, `split`, `substring`, `substring_index`, `translate`, `trim`, `unbase64`, `upper`, and other commonly used string functions.

Window

Spark SQL supports window functions for analytics. A window function performs a calculation across a set of rows that are related to the current row. The built-in window functions provided by Spark SQL include `cumeDist`, `denseRank`, `lag`, `lead`, `ntile`, `percentRank`, `rank`, *and* `rowNumber`.

UDFs and UDAFs

Spark SQL allows user-defined functions (UDFs) and user-defined aggregation functions (UDAFs). Both UDFs and UDAFs perform custom computations on a dataset. A UDF performs custom computation one row at a time and returns a value for each row. A UDAF applies custom aggregation over groups of rows.

UDFs and UDAFs can be used just like the built-in functions after they have been registered with Spark SQL.

Interactive Analysis Example

As you have seen in previous sections, Spark SQL can be used as an interactive data analysis tool. The previous sections used toy examples so that they were easy to understand. This section shows how Spark SQL can be used for interactive data analysis on a real-world dataset. You can use either language integrated queries or SQL/HiveQL for data analysis. This section demonstrates how they can be used interchangeably.

The dataset that will be used is the Yelp challenge dataset. You can download it from `www.yelp.com/dataset_challenge`. It includes a number of data files. The example uses the `yelp_academic_dataset_business.json` file, which contains information about businesses. It contains a JSON object per line. Each JSON object contains information about a business, including its name, city, state, reviews, average rating, category and other attributes. You can learn more details about the dataset on Yelp's website.

Let's launch the Spark shell from a terminal, if it is not already running.

```
path/to/spark/bin/spark-shell --master local[*]
```

As discussed previously, the --master argument specifies the Spark master to which the Spark shell should connect. For the examples in this section, you can use Spark in local mode. If you have a real Spark cluster, change the master URL argument to point to your Spark master. The rest of the code in this section does not require any change.

For readability, I have split some of the example code statements into multiple lines. However, the Spark shell executes a statement as soon as you press the *ENTER* key. For multiline code statements, you can use Spark shell's paste mode (:paste). Alternatively, enter a complete statement on a single line.

Consider the following example.

```
biz.filter("...")
   .select("...")
   .show()
```

In the Spark shell, type it without the line breaks, as shown next.

```
biz.filter("...").select("...").show()
```

Consider another example.

```
sqlContext.sql("SELECT x, count(y) as total FROM t
                GROUP BY x
                ORDER BY total")
          .show(50)
```

Type the preceding code in the Spark shell without line breaks, as shown next.

```
sqlContext.sql("SELECT x, count(y) as total FROM t GROUP BY x ORDER BY total ").show(50)
```

Let's get started now.

Since you will be using a few classes and functions from the Spark SQL library, you need the following import statement.

```
import org.apache.spark.sql._
```

Let's create a DataFrame from the Yelp businesses dataset.

```
val biz = sqlContext.read.json("path/to/yelp_academic_dataset_business.json")
```

The preceding statement is equivalent to this:

```
val biz = sqlContext.read.format("json").load("path/to/yelp_academic_dataset_business.json")
```

Make sure that you specify the correct file path. Replace "path/to" with the name of the directory where you unpacked the Yelp dataset. Spark SQL will throw an exception if it cannot find the file at the specified path.

Spark SQL reads the entire dataset once to infer the schema from a JSON file. Let's check the schema inferred by Spark SQL.

```
biz.printSchema()
```

Partial output is shown next.

```
root
 |-- attributes: struct (nullable = true)
 |     |-- Accepts Credit Cards: string (nullable = true)
 |     |-- Ages Allowed: string (nullable = true)
 |     |-- Alcohol: string (nullable = true)
 ...
 ...
 ...
 |-- name: string (nullable = true)
 |-- open: boolean (nullable = true)
 |-- review_count: long (nullable = true)
 |-- stars: double (nullable = true)
 |-- city: string (nullable = true)
 |-- state: string (nullable = true)
 |-- type: string (nullable = true)
```

To use the SQL/HiveQL interface, you need to register the biz DataFrame as a temporary table.

```
biz.registerTempTable("biz")
```

Now you can analyze the Yelp businesses dataset using either the DataFrame API or SQL/HiveQL. I will show the code for both, but show the output only for the SQL/HiveQL queries.

Let's cache the data in memory since you will be querying it more than one once.

Language-Integrated Query

```
biz.cache()
```

SQL

```
sqlContext.cacheTable("biz")
```

You can also cache a table using the CACHE TABLE statement, as shown next.

```
sqlContext.sql("CACHE TABLE biz")
```

Note that unlike RDD caching, Spark SQL immediately caches a table when you use the CACHE TABLE statement.

Since the goal of this section is to just demonstrate how you can use the Spark SQL library for interactive data analysis, the steps from this point onward have no specific ordering. I chose a few seemingly random queries. You may analyze data in a different sequence. Before each example, I will state the goal and then show the Spark SQL code for achieving that goal.

Let's find the number of businesses in the dataset.

Language-Integrated Query

```
val count = biz.count()
```

SQL

```
sqlContext.sql("SELECT count(1) as businesses FROM biz").show
```

```
+----------+
|businesses|
+----------+
|     61184|
+----------+
```

Next, let's find the count of businesses by state.

Language-Integrated Query

```
val bizCountByState = biz.groupBy("state").count
bizCountByState.show(50)
```

SQL

```
sqlContext.sql("SELECT state, count(1) as businesses FROM biz GROUP BY state").show(50)
```

```
+-----+----------+
|state|businesses|
+-----+----------+
|  XGL|         1|
|   NC|      4963|
|   NV|     16485|
|   AZ|     25230|
...

...
|  MLN|       123|
|  NTH|         1|
|   MN|         1|
+-----+----------+
```

Next, let's find the count of businesses by state and sort the result by count in descending order.

Language-Integrated Query

```
val resultDF = biz.groupBy("state").count
resultDF.orderBy($"count".desc).show(5)
```

SQL

```
sqlContext.sql("SELECT state, count(1) as businesses FROM biz GROUP BY state ORDER BY
businesses DESC").show(5)
```

```
+-----+------------+
|state|businesses  |
+-----+------------+
|   AZ|       25230|
|   NV|       16485|
|   NC|        4963|
|   QC|        3921|
|   PA|        3041|
+-----+------------+
only showing top 5 rows
```

You could have also written the language integrated query version as follows:

```
val resultDF = biz.groupBy("state").count
resultDF.orderBy(resultDF("count").desc).show(5)
```

The difference between the first and second version is how you refer to the *count* column in the resultDF DataFrame. You can think of $"count" as a shortcut for resultDF("count"). If a method or a function expects an argument of type Column, you can use either syntax.

Next, let's find five businesses with a five-star rating.

Language-Integrated Query

```
biz.filter(biz("stars") <=> 5.0)
   .select("name","stars", "review_count", "city",  "state")
   .show(5)
```

The preceding code first uses the filter method to filter the businesses that have average rating of 5.0. The <=> operator does equality test that is safe for null values. After filtering the businesses, it selects a subset of the columns that are of interest. Finally, it displays five businesses on the console.

SQL

```
sqlContext.sql("SELECT name, stars, review_count, city, state FROM biz WHERE stars=5.0").show(5)
```

```
+--------------------+-----+------------+------------+-----+
|                name|stars|review_count|        city|state|
+--------------------+-----+------------+------------+-----+
|   Alteration World|  5.0|           5|    Carnegie|   PA|
|American Buyers D...|  5.0|           3|   Homestead|   PA|
|Hunan Wok Chinese...|  5.0|           4|West Mifflin|   PA|
|      Minerva Bakery|  5.0|           7|  McKeesport|   PA|
|                Vivo|  5.0|           3|    Bellevue|   PA|
+--------------------+-----+------------+------------+-----+
only showing top 5 rows
```

For the next example, let's find three businesses with five stars in Nevada.

Language Integrated Query

```
biz.filter($"stars" <=> 5.0 && $"state" <=> "NV")
  .select("name","stars", "review_count", "city",  "state")
  .show(3)
```

SQL

```
sqlContext.sql("SELECT name, stars, review_count, city, state FROM biz WHERE state = 'NV'
AND stars = 5.0").show(3)
```

```
+--------------------+-----+------------+---------+-----+
|                name|stars|review_count|     city|state|
+--------------------+-----+------------+---------+-----+
|              Adiamo|  5.0|           4|Henderson|   NV|
|CD Young's Profes...|  5.0|           8|Henderson|   NV|
|Liaisons Salon & Spa|  5.0|           5|Henderson|   NV|
+--------------------+-----+------------+---------+-----+
only showing top 3 rows
```

Next, let's find the total number of reviews in each state.

Language Integrated Query

```
biz.groupBy("state").sum("review_count").show()
```

SQL

```
sqlContext.sql("SELECT state, sum(review_count) as reviews FROM biz GROUP BY state").show()
```

```
+-----+-------+
|state|reviews|
+-----+-------+
|  XGL|      3|
|   NC| 102495|
|   NV| 752904|
|   AZ| 636779|
...
...
|   QC|  54569|
|  KHL|      8|
|   RP|     75|
+-----+-------+
only showing top 20 rows
```

Next, let's find count of businesses by stars.

Language Integrated Query

```
biz.groupBy("stars").count.show()
```

SQL

```
sqlContext.sql("SELECT stars, count(1) as businesses FROM biz GROUP BY stars").show()
```

```
+-----+----------+
|stars|businesses|
+-----+----------+
|  1.0|       637|
|  3.5|     13171|
|  4.5|      9542|
|  3.0|      8335|
|  1.5|      1095|
|  5.0|      7354|
|  2.5|      5211|
|  4.0|     13475|
|  2.0|      2364|
+-----+----------+
```

Next, let's find the average number of reviews for a business by state.

Language-Integrated Query

```
val avgReviewsByState = biz.groupBy("state").avg("review_count")
avgReviewsByState.show()
```

SQL

```
sqlContext.sql("SELECT state, AVG(review_count) as avg_reviews FROM biz GROUP BY state").show()
```

```
+-----+------------------+
|state|       avg_reviews|
+-----+------------------+
|  XGL|               3.0|
|   NC|20.651823493854522|
|   NV| 45.67206551410373|
|   AZ|25.238961553705906|
...
...
|   QC|13.917112981382301|
|  KHL|               8.0|
|   RP| 5.769230769230769|
+-----+------------------+
only showing top 20 rows
```

Next, let's find the top five states by the average number of reviews for a business.

Language Integrated Query

```
biz.groupBy("state")
    .avg("review_count")
    .withColumnRenamed("AVG(review_count)", "rc")
    .orderBy($"rc".desc)
    .selectExpr("state", "ROUND(rc) as avg_reviews")
    .show(5)
```

SQL

```
sqlContext.sql("SELECT state, ROUND(AVG(review_count)) as avg_reviews FROM biz GROUP BY
state ORDER BY avg_reviews DESC LIMIT 5").show()
```

```
+-----+-----------+
|state|avg_reviews|
+-----+-----------+
|   NV|       46.0|
|   AZ|       25.0|
|   PA|       24.0|
|   NC|       21.0|
|   IL|       21.0|
+-----+-----------+
```

Next, let's find the top 5 businesses in Las Vegas by average stars and review counts.

Language Integrated Query

```
biz.filter($"city" === "Las Vegas")
    .sort($"stars".desc, $"review_count".desc)
    .select($"name", $"stars", $"review_count")
    .show(5)
```

SQL

```
sqlContext.sql("SELECT name, stars, review_count FROM biz WHERE city = 'Las Vegas' ORDER BY
stars DESC, review_count DESC LIMIT 5 ").show
```

```
+--------------------+-----+------------+
|                name|stars|review_count|
+--------------------+-----+------------+
|      Art of Flavors|  5.0|         321|
|Free Vegas Club P...|  5.0|         285|
|Fabulous Eyebrow ...|  5.0|         244|
|Raiding The Rock ...|  5.0|         199|
|            Eco-Tint|  5.0|         193|
+--------------------+-----+------------+
```

Next, let's write the data in Parquet format.

```
biz.write.mode("overwrite").parquet("path/to/yelp_business.parquet")
```

You can read the Parquet files created in the previous step, as shown next.

```
val ybDF = sqlContext.read.parquet("path/to/yelp_business.parquet")
```

Interactive Analysis with Spark SQL JDBC Server

This section shows how you can explore the Yelp dataset using just SQL/HiveQL. Scala is not used at all. For this analysis, you will use the Spark SQL Thrift/JDBC/ODBC server and the Beeline client. Both come prepackaged with Spark.

The first step is to launch the Spark SQL Thrift/JDBC/ODBC server from a terminal. Spark's sbin directory contains a script for launching it.

```
path/to/spark/sbin/start-thriftserver.sh --master local[*]
```

The second step is to launch Beeline, which is a CLI (command line interface) client for Spark SQL Thrift/JDBC server. Conceptually, it is similar to the mysql client for MySQL or psql client for PostgreSQL. It allows you to type a HiveQL query, sends the typed query to a Spark SQL Thrift/JDBC server for execution, and displays the results on the console.

Spark's bin directory contains a script for launching Beeline. Let's open another terminal and launch Beeline.

```
path/to/spark/bin/beeline
```

You should be now inside the Beeline shell and see the Beeline prompt, as shown next.

```
Beeline version 1.5.2 by Apache Hive
beeline>
```

The third step is to connect to the Spark SQL Thrift/JDBC server from the Beeline shell.

```
beeline>!connect jdbc:hive2://localhost:10000
```

The connect command requires a JDBC URL. The default JDBC port for the Spark SQL Thrift/JDBC server is 10000.

Beeline will ask for a username and password. Enter the username that you use to login into your system and a blank password.

```
beeline> !connect jdbc:hive2://localhost:10000
```

```
scan complete in 26ms
Connecting to jdbc:hive2://localhost:10000
Enter username for jdbc:hive2://localhost:10000: your-user-name
Enter password for jdbc:hive2://localhost:10000:
Connected to: Spark SQL (version 1.5.2)
Driver: Spark Project Core (version 1.5.2)
Transaction isolation: TRANSACTION_REPEATABLE_READ
0: jdbc:hive2://localhost:10000>
```

At this point, you have an active connection between the Beeline client and Spark SQL server.

However, the Spark SQL server does not yet know about the Yelp dataset. Let's create a temporary table that points to the Yelp dataset.

```
0: jdbc:hive2://localhost:10000> CREATE TEMPORARY TABLE biz USING org.apache.spark.sql.json
OPTIONS (path "path/to/yelp_academic_dataset_business.json");
```

Make sure to replace "*path/to*" with the path for the directory where you unpacked the Yelp dataset on your machine. Spark SQL will throw an exception if it cannot find the file at the specified path.

The CREATE TEMPORARY TABLE command creates an external temporary table in Hive metastore. A temporary table exists only while the Spark SQL JDBC server is running.

Now you can analyze the Yelp dataset using just plain SQL/HiveQL queries. A few examples are shown next.

```
0: jdbc:hive2://localhost:10000> SHOW TABLES;
```

```
+------------+--------------+--+
| tableName  | isTemporary  |
+------------+--------------+--+
| biz        | true         |
+------------+--------------+--+
```

```
0: jdbc:hive2://localhost:10000> SELECT count(1) from biz;
```

```
+--------+--+
|  _c0   |
+--------+--+
| 61184  |
+--------+--+
```

0: jdbc:hive2://localhost:10000> SELECT state, count(1) as cnt FROM biz GROUP BY state ORDER BY cnt DESC LIMIT 5;

```
+--------+--------+--+
| state  |  cnt   |  |
+--------+--------+--+
| AZ     | 25230  |  |
| NV     | 16485  |  |
| NC     | 4963   |  |
| QC     | 3921   |  |
| PA     | 3041   |  |
+--------+--------+--+
5 rows selected (3.241 seconds)
```

0: jdbc:hive2://localhost:10000> SELECT state, count(1) as businesses, sum(review_count) as reviews FROM biz GROUP BY state ORDER BY businesses DESC LIMIT 5;

```
+--------+------------+----------+--+
| state  | businesses |  reviews |  |
+--------+------------+----------+--+
| AZ     | 25230      | 636779   |  |
| NV     | 16485      | 752904   |  |
| NC     | 4963       | 102495   |  |
| QC     | 3921       | 54569    |  |
| PA     | 3041       | 72409    |  |
+--------+------------+----------+--+
5 rows selected (1.293 seconds)
```

0: jdbc:hive2://localhost:10000> SELECT name, review_count, stars, city, state from biz WHERE stars = 5.0 ORDER BY review_count DESC LIMIT 5;

```
+---------------------------+--------------+--------+-------------+--------+--+
|           name            | review_count | stars  |    city     | state  |  |
+---------------------------+--------------+--------+-------------+--------+--+
| Art of Flavors            | 321          | 5.0    | Las Vegas   | NV     |  |
| PNC Park                  | 306          | 5.0    | Pittsburgh  | PA     |  |
| Gaucho Parrilla Argentina | 286          | 5.0    | Pittsburgh  | PA     |  |
| Free Vegas Club Passes    | 285          | 5.0    | Las Vegas   | NV     |  |
| Little Miss BBQ           | 267          | 5.0    | Phoenix     | AZ     |  |
+---------------------------+--------------+--------+-------------+--------+--+
5 rows selected (0.511 seconds)
```

Summary

Spark SQL is a Spark library that makes it simple to do fast analysis of structured data. It provides more than just a SQL interface to Spark. It improves Spark usability, developer productivity, and application performance.

Spark SQL provides a unified interface for processing structured data from a variety of data sources. It can be used to process data stored on a local file system, HDFS, S3, JDBC-compliant databases, and NoSQL datastores. It allows you to process data from any of these data sources using SQL, HiveQL, or the DataFrame API.

Spark SQL also provides a distributed SQL query server that supports Thrift, JDBC and ODBC clients. It allows you to interactively analyze data using just SQL/HiveQL. The Spark SQL Thrift/JDBC/ODBC server can be used with any third-party BI or data visualization application that supports the JDBC or ODBC interface.

CHAPTER 8

■ ■ ■

Machine Learning with Spark

Interest in machine learning is growing by leaps and bounds. It has gained a lot of momentum in recent years for a few reasons. The first reason is performance improvements in hardware and algorithms. Machine learning is compute-intensive. With proliferation of multi-CPU and multi-core machines and efficient algorithms, it has become feasible to do machine learning computations in reasonable time. The second reason is that machine learning software has become freely available. Many good quality open source machine learning software are available now for anyone to download. The third reason is that MOOCs (massive open online courses) have created tremendous awareness about machine learning. These courses have democratized the knowledge required to use machine learning. Machine learning skills are no longer limited to a few people with Ph.D. in Statistics. Anyone can now learn and apply machine learning techniques.

Machine learning is embedded in many applications that we use on a daily basis. Apple, Google, Facebook, Twitter, LinkedIn, Amazon, Microsoft, and numerous others use machine learning under the hood in many of their products. Examples of machine learning applications include driverless car, motion-sensing game consoles, medical diagnosis, email spam filtering, image recognition, voice recognition, fraud detection, and movie, song, and book recommendations.

This chapter introduces core machine learning concepts and then discusses the machine learning libraries that Spark provides. Machine learning is a broad topic; it can be the subject of a book by itself. In fact, several books have been written on it. Covering it in detail is out of scope for this book; the next section introduces the basic concepts, so that you can follow the rest of the material in this chapter. If you are familiar with machine learning, you can skip the next section.

Introducing Machine Learning

Machine learning is the science of training a system to learn from data and act. The logic that drives the behavior of a machine learning-based system is not explicitly programmed but learnt from data.

A simple analogy is an infant learning to speak by observing others. Babies are not born with any language skills, but they learn to understand and speak words by observing others. Similarly, in machine learning, we train a system with data instead of explicitly programming its behavior.

To be specific, a machine learning algorithm infers patterns and relationships between different variables in a dataset. It then uses that knowledge to generalize beyond the training dataset. In other words, a machine learning algorithm learns to predict from data.

Let's introduce some of the terms generally used in the context of machine learning.

Features

A feature represents an attribute or a property of an observation. It is also called a variable. To be more specific, a feature represents an independent variable.

In a tabular dataset, a row represents an observation and column represents a feature. For example, consider a tabular dataset containing user profiles, which includes fields such as age, gender, profession, city, and income. Each field in this dataset is a feature in the context of machine learning. Each row containing a user profile is an observation.

Features are also collectively referred to as dimensions. Thus, a dataset with high dimensionality has a large number of features.

Categorical Features

A categorical feature or variable is a descriptive feature. It can take on one of a fixed number of discrete values. It represents a qualitative value, which is a name or a label.

The values of a categorical feature have no ordering. For example, in the user profile dataset mentioned earlier, gender is a categorical feature. It can take on only one of two values, each of which is a label. In the same dataset, profession is also a categorical variable, but it can take on one of several hundred values.

Numerical Features

A numerical feature or variable is a quantitative variable that can take on any numerical value. It describes a measurable quantity as a number. The values in a numerical feature have mathematical ordering. For example, in the user profile dataset mentioned earlier, income is a numerical feature.

Numerical features can be further classified into discrete and continuous features. A discrete numerical feature can take on only certain values. For example, the number of bedrooms in a home is a discrete numerical feature. A continuous numerical feature can take on any value within a finite or infinite interval. An example of a continuous numerical feature is temperature.

Labels

A label is a variable that a machine learning system learns to predict. It is the dependent variable in a dataset. Labels can be a classified into two broad categories: categorical and numerical.

A categorical label represents a class or category. For example, for a machine learning application that classifies news articles into different categories such as politics, business, technology, sports, or entertainment, the category of a news article is a categorical label.

A numerical label is a numerical dependent variable. For example, for a machine learning application that predicts the price of a house, price is a numerical label.

Models

A model is a mathematical construct for capturing patterns within a dataset. It estimates the relationship between the dependent and independent variables in a dataset. It has predictive capability. Given the values of the independent variables, it can calculate or predict the value for the dependent variable. For example, consider an application that forecasts quarterly sales for a company. The independent variables are number of sales people, historical sales, macro-economic conditions, and other factors. Using machine learning, a model can be trained to predict quarterly sales for any given combination of these factors.

A model is basically a mathematical function that takes features as input and outputs a value. It can be represented in software in numerous ways. For example, it can be represented by an instance of a class. We will see a few concrete examples later in this chapter.

A model along with a machine learning algorithm forms the heart of a machine learning system. A machine learning algorithm trains a model with data; it fits a model over a dataset, so that the model can predict the label for a new observation.

Training a model is a compute intensive task, while using it is not as compute intensive. A model is generally saved to disk, so that it can be used in future without having to go through the compute intensive training step again. A serialized model can also be shared with other applications. For example, a machine learning system may consist of two applications, one that trains a model and another that uses a model.

Training Data

The data used by a machine learning algorithm to train a model is called training data or training set. It is historical or known data. For example, a spam filtering algorithm uses a known set of spam and non-spam emails.

The training data can be classified into two categories: labeled and unlabeled.

Labeled

A labeled dataset has a label for each observation. One of the columns in the dataset contains the labels. For example, a database of homes sold in the last ten years is a labeled dataset for a machine learning application that predicts the price of a home. The label in this case is home price, which is known for homes sold in the past. Similarly, a spam filtering application is trained with a large dataset of emails, some of which are labeled as spam and others as non-spam.

Unlabeled

An unlabeled dataset does not have a column that can be used as a label. For example, consider a transaction database for an e-commerce site. It records all the online purchases made through that site. This database does not have a column that indicates whether a transaction was normal or fraudulent. So for fraud detection purposes, this is an unlabeled dataset.

Test Data

The data used for evaluating the predictive performance of a model is called test data or test set. After a model has been trained, its predictive capabilities should be tested on a known dataset before it is used on new data.

Test data should be set aside before training a model. It should not be used at all during the training phase; it should not be used for training or optimizing a model. In fact, it should not influence the training phase in any manner; do not even look at it during the training phase. A corollary to this is that a model should not be tested with the training dataset. It will perform very well on the observations from the training set. It should be tested on data that was not used in training it.

Generally, a small proportion of a dataset is held out for testing before training a model. The exact percentage depends on a number of factors such as the size of a dataset and the number of independent variables. A general rule of thumb is to use 80% of data for training a model and set aside 20% as test data.

Machine Learning Applications

Machine learning is used for a variety of tasks in different fields. A large number of applications use machine learning, and that number is increasing every day. The machine learning tasks can be broadly grouped into the following categories:

- Classification

- Regression

- Clustering

- Anomaly detection

- Recommendation

- Dimensionality reduction

Classification

The goal while solving a classification problem is to predict a class or category for an observation. A class is represented by a label. The labels for the observations in the training dataset are known, and the goal is to train a model that predicts the label for a new unlabeled observation. Mathematically, in a classification task, a model predicts the value of a categorical variable.

Classification is a common task in many fields. For example, spam filtering is a classification task. The goal of a spam filtering system is to classify an email as a spam or not. Similarly, tumor diagnosis can be treated as a classification problem. A tumor can be benign or cancerous. The goal in this case is to predict whether a tumor is benign or cancerous. Another example of a classification task is determining credit risk of a borrower. Using information such as an individual's income, outstanding debt, and net worth, a credit rating is assigned to an individual.

Machine learning can be used for both binary and multi-class classification. The previous paragraph described a few examples of binary classification. In binary classification, the observations in a dataset can be grouped into two mutually exclusive classes. Each observation or sample is either a positive or negative example.

In multi-class classification, the observations in a dataset can be grouped into more than two classes. For example, handwritten zip-code recognition is a multi-class classification problem with ten classes. In this case, the goal is to detect whether a handwritten character is one of the digits between 0-9. Each digit represents a class. Similarly, image recognition is a multi-class classification task, which has many applications. One of the well-known applications is a self-driving or driver-less car. Another application is Xbox Kinect360, which infers body parts and position using machine learning.

Regression

The goal while solving a regression problem is to predict a numerical label for an unlabeled observation. The numerical labels are known for the observations in the training dataset and a model is trained to predict the label for a new observation.

Examples of regression tasks include home valuation, asset trading, and forecasting. In home valuation, the value of a home is the numerical variable that a model predicts. In asset trading, regression techniques are used to predict the value of an asset such as a stock, bond, or currency. Similarly, sales or inventory forecasting is a regression task.

Clustering

In clustering, a dataset is split into a specified number of clusters or segments. Elements in the same cluster are more similar to each other than to those in other clusters. The number of clusters depends on the application. For example, an insurance company may segment its customers into three clusters: low-risk, medium-risk and high-risk. On the other hand, an application may segment users on a social network into 10 communities for research purposes.

Some people find clustering confusingly similar to classification. They are different. In a classification task, a machine learning algorithm trains a model with a labeled dataset. Clustering is used with unlabeled datasets. In addition, although a clustering algorithm splits a dataset into a specified number of clusters, it does not assign a label to any cluster. A user has to determine what each cluster represents.

A popular example of clustering is customer segmentation. Organizations use clustering as a data-driven technique for creating customer segments, which can be targeted with different marketing programs.

Anomaly Detection

In anomaly detection, the goal is to find outliers in a dataset. The underlying assumption is that an outlier represents an anomalous observation. Anomaly detection algorithms are used with unlabeled data.

Anomaly detection has many applications in different fields. In manufacturing, it is used for automatically finding defective products. In data centers, it is used for detecting bad systems. Websites use it for fraud detection. Another common use-case is detecting security attacks. Network traffic associated with a security attack is unlike normal network traffic. Similarly, hacker activity on a machine will be different from a normal user activity.

Recommendation

The goal of a recommendation system, also known as recommender system, is to recommend a product to a user. It learns from users' past behavior to determine user preferences. A user rates different products, and over time, a recommendation system learns this user's preferences. In some cases, a user may not explicitly rate a product but provide implicit feedback through actions such as purchase, click, view, like, or share.

A recommendation system is one of the well-known examples of machine learning. It is getting embedded in more and more applications. Recommendation systems are used to recommend news articles, movies, TV shows, songs, books, and other products. For example, Netflix uses recommender systems to recommend movies and shows to its subscribers. Similarly, Spotify, Pandora, and Apple use recommendation systems to recommend songs to their subscribers.

The two commonly used techniques for building recommendation systems are collaborative filtering and content-based recommendation. In collaborative filtering, the properties of a product or user preferences are not explicitly programmed. The algorithm assumes that the user preferences and products have latent features, which it automatically learns from ratings of different products by different users. The input dataset is in a tabular format, where each row contains only a user id, product id, and rating. Collaborative filtering learns latent user and product feature just from these three fields. It learns users with similar preferences and products with similar properties. The trained model can then be used to recommend products to a user. The products recommended to a user are those rated highly by other users with similar preferences.

A content-based recommendation system uses explicitly specified product properties to determine product similarity and make recommendations. For example, a movie has properties such as genre, lead actor, director, and year released. In a content-based system, every movie in a movie database will have these properties recorded. For a user who mostly watches comedy movies, a content-based system will recommend a movie having genre as comedy.

Dimensionality Reduction

Dimensionality reduction is a useful technique for reducing the cost and time it takes to train a machine learning system. Machine learning is a compute intensive task. The computation complexity and cost increases with the number of features or dimensions in a dataset. The goal in dimensionality reduction is to reduce the number of features in a dataset without significantly impacting the predictive performance of a model.

A dataset may have so many dimensions that it is prohibitively expensive to use it for machine learning. For example, a dataset may have several thousand features. It may take days or weeks to train a system with this dataset. With dimensionality reduction techniques, it can be used to train a machine learning system in a more reasonable time.

The basic idea behind dimensionality reduction is that a dataset may have several features that have low or zero predictive power. A dimensionality reduction algorithm automatically eliminates these features from a dataset. Only the features with most predictive power are used for machine learning. Thus, dimensionality reduction techniques reduce the computational complexity and cost of machine learning.

Machine Learning Algorithms

Machine learning algorithms use data to train a model. The process of training a model is also referred to as fitting a model with data. In other words, a machine learning algorithm fits a model on a training dataset.

Depending on the type of the training data, machine learning algorithms are broadly grouped into two categories: supervised and unsupervised machine learning.

Supervised Machine Learning Algorithms

A supervised machine learning algorithm trains a model with a labeled dataset. It can be used only with labeled training datasets.

Each observation in the training dataset has a set of features and a label. The dependent variable, also known as the response variable, represents the label. The independent variables, also known as explanatory or predictor variables, represent the features. A supervised machine learning algorithm learns from data to estimate or approximate the relationship between a response variable and one or more predictor variables.

The labels in a training dataset may be generated manually or sourced from another system. For example, for spam filtering, a large sample of emails are collected and manually labeled as spam or not. On the other hand, for sales forecasting, label will be historical sales, which can be sourced from a sales database.

Supervised machine learning algorithms can be broadly grouped into two categories: regression and classification algorithms.

Regression Algorithms

A regression algorithm trains a model with a dataset that has a numerical label. The trained model can then predict numerical labels for new unlabeled observations.

Depending on the number of predictor and response variables, regression tasks can be grouped in three categories: simple, multiple and multivariate regression. Simple regression involves one response and one predictor variable. Multiple regression involves one response and multiple predictor variables. Multivariate regression involves several response and several predictor variables.

The commonly used supervised machine learning algorithms for regression tasks include linear regression, decision trees, and ensembles of trees.

Linear Regression

Linear regression algorithms fit a linear model with coefficients using training data. A linear model is a linear combination of a set of coefficients and explanatory variables. The algorithm estimates the unknown coefficients, also known as model parameters, from training data. The fitted coefficients minimize the sum of the squares of the difference between predicted and actual observed labels in the training dataset.

A simple example of a linear model is shown here.

$$y = \theta_0 + \theta_1 x_1 + \theta_2 x_2 + \theta_3 x_1 x_2$$

In this equation, y is the label or dependent variable and x_1 and x_2 are the features or independent variables. The values for y, x_1 and x_2 are known for each observation in a training set. A linear regression algorithm estimates the values for θ_0, θ_1, θ_2 and θ_3 using the observations in the training set. With the values of θ_0, θ_1, θ_2 and θ_3 known, the value for y can be calculated for any values of x_1 and x_2.

A graphical representation of a linear regression function is shown in Figure 8-1.

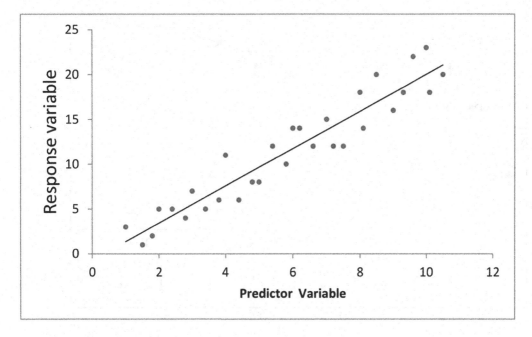

Figure 8-1. *Linear regression*

Linear regression is one of the simplest and oldest machine learning algorithms. It has been researched rigorously and used in many applications. It is used in forecasting, planning, and asset valuation applications. For example, it can be used to forecast sales. It can be used to forecast system usage in a datacenter environment. It can be used to determine the value of a home.

Isotonic Regression

The Isotonic Regression algorithm fits a non-decreasing function to a training dataset. It finds the best least squares fit to a training dataset with the constraint that the trained model must be a non-decreasing function. A least squares function minimizes the sum of the squares of the difference between predicted and actual labels in the training dataset. Unlike linear regression, the Isotonic Regression algorithm does not assume any form for the target function such as linearity.

The diagram in Figure 8-2 compares a model trained with Isotonic Regression algorithm to that trained with a linear regression algorithm.

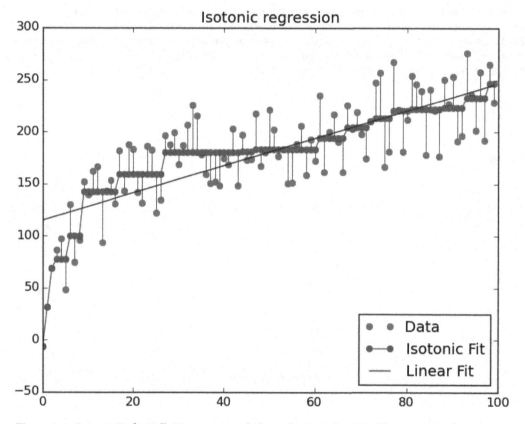

Figure 8-2. *Isotonic Fit (© Nelle Varoquaux and Alexandre Gramfort. Used by permission.)*

Decision Trees

The decision tree algorithm infers a set of decision rules from a training dataset. Essentially, it creates a decision tree (see Figure 8-3) that can be used to predict the numeric label for an observation.

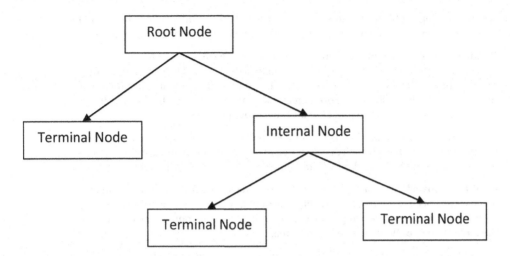

Figure 8-3. *Decision tree*

A tree is a hierarchal collection of nodes and edges. Unlike a graph, there are no loops in a tree. A non-leaf node is called an internal or split node. A leaf node is called a terminal node.

In a decision tree, each internal node tests the value of a feature or predictor variable. The observations in the training dataset are segmented into a number of regions using these tests. A leaf node represents a region and stores the average value of all the observations belonging to a region.

Given a new unlabeled observation, a decision tree model starts at the root node and evaluates the observation features against the internal nodes. It traverses down a tree until it arrives at a terminal node. The value stored at the matching terminal node is returned as the predicted label. Thus, a decision tree model conceptually implements hierarchal if-else statements. It performs a series of tests on the features to predict a label.

Decision trees can be used for both regression and classification tasks. The "Classification Algorithms" section describes how decision trees can be used for classification tasks.

The decision tree algorithm has many advantages over other more sophisticated machine learning algorithms. First, models trained by a decision tree are easy to understand and explain. Second, it can easily handle both categorical and numerical features. Third, it requires little data preparation. For example, unlike other algorithms, it does not require feature scaling.

Ensembles

Although the Decision Tree algorithm has many advantages, it also has one big disadvantage. Decision trees do not have the same level of predictive accuracy as models trained with algorithms that are more sophisticated. This problem is addressed by using a collection of decision trees instead of just one decision tree.

The machine learning algorithms that combine multiple models to generate a more powerful model are called ensemble learning algorithms. A model trained by an ensemble learning algorithm combines the predictions of several base models to improve generalizability and predictive accuracy over a single model. Ensemble learning algorithms are among the top performers for classification and regression tasks.

The commonly used ensemble algorithms include Random Forests and Gradient-Boosted Trees. Both algorithms train an ensemble of decision trees, but use different techniques for growing the trees.

- **Random Forests.** The Random Forest algorithm trains each decision tree in an ensemble independently using a random sample of data. In addition, each decision tree is trained using a subset of the features. The number of trees in an ensemble is of the order of hundreds. Random Forest creates an ensemble model that has a better predictive performance than that of a single decision tree model.

  For a regression task, a Random Forest model takes an input feature vector and gets a prediction from each decision tree in the ensemble. It averages the numeric labels return by all the trees and returns the average as its prediction.

- **Gradient-Boosted Trees.** The Gradient-Boosted Trees (GBTs) algorithm also trains an ensemble of decision trees. However, it sequentially trains each decision tree. It optimizes each new tree using information from previously trained trees. Thus, the model becomes better with each new tree.

  GBT can take longer to train a model since it trains one tree at a time. In addition, it is prone to overfitting if a large number of trees are used in an ensemble. However, each tree in a GBT ensemble can be shallow, which are faster to train.

Classification Algorithms

Classification algorithms train models that predict categorical values. The dependent or response variable in the training dataset is a categorical variable. In other words, the label is a categorical variable.

The model trained by a classification algorithm can be a binary, multi-class, or multi-label classifier.

A binary classifier classifies observations into two categories: positive or negative. The predicted label has only two classes.

A multi-class classifier predicts a label that can have more than two classes. For example, a multi-class classifier can be used to classify images of animals. The label in this example can be cat, dog, hamster, lion, or some other animal.

A multi-label classifier can output more than one label for the same observation. For example, a classifier that categorizes news articles can output more than one label for an article that is related to both sports and business.

The commonly used supervised machine learning algorithms for classification tasks include the following algorithms.

- Logistic Regression

- Support Vector Machine (SVM)

- Naïve Bayes

- Decision Trees

- Ensembles

- Neural Network

Logistic Regression

The logistic regression algorithm trains a linear model that can be used for classification tasks. Specifically, the generated model can be used for predicting the probability of occurrence of an event.

Logistic regression uses a logistic or sigmoid function to model the probabilities for the possible labels of an unlabeled observation. An example of a logistic function is shown next.

$$P(Y=1\,|\,x;\theta) = \frac{1}{1+e^{-(\theta_0+\theta_1 x_1+\theta_2 x_2)}}$$

In the preceding equation, x_1, x_2, and x_3 represent predictor variables and Y represents a class for an observation. The logistic regression algorithm estimates the value for θ_0, θ_1, and θ_2 using a training dataset. After the values for θ_0, θ_1, and θ_2 have been learned, the preceding equation can be used to calculate the probability of a label class for a given observation.

A graphical representation of a logistic function, also known as a *sigmoid function*, is shown in Figure 8-4.

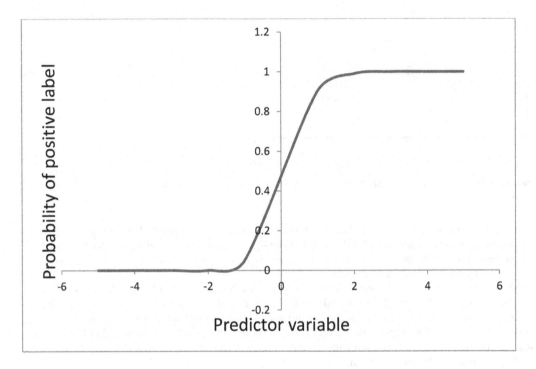

Figure 8-4. A logistic regression curve

Support Vector Machine (SVM)

The Support Vector Machine (SVM) algorithm trains an optimal classifier. Conceptually, it learns from a training dataset an optimal hyperplane (see Figure 8-5) for classifying a dataset. It finds the best hyperplane that separates training observations of one class from those of the other class. The support vectors are the feature vectors that are closest to the separating hyperplane.

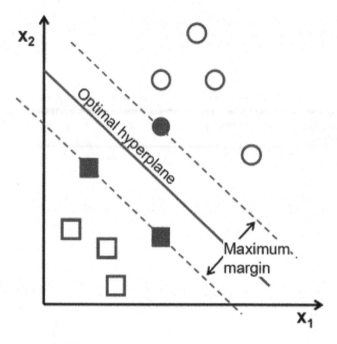

Figure 8-5. *SVM classifier (Source: docs.opencv.org/2.4/doc/tutorials/ml/introduction_to_svm/introduction_to_svm.html)*

The best hyperplane is the one with the largest margin between two classes of observations. Margin in this context is the width of a slab that cleanly separates the observations in the training set. In other words, the margin between the separating hyperplane and the nearest feature vectors from both classes is maximal. The diagram in Figure 8-5 illustrates this point.

SVM can be used as a kernel-based method. A kernel-based method implicitly maps feature vectors into a higher-dimensional space where it is easier to find an optimal hyperplane for classifying observations (see Figure 8-6). For example, it may be difficult to find a hyperplane that separates positive and negative examples in a two-dimensional space. However, if the same data is mapped to a three or higher dimensional space, it may be easier to find a hyperplane that cleanly separates the positive and negative observations. The next diagram illustrates this approach.

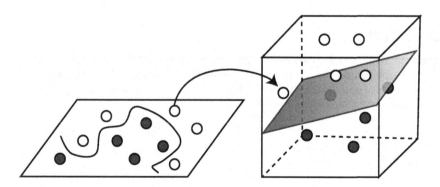

Figure 8-6. *Kernel method*

A kernel based method uses a kernel function, which is a similarity function. The kernel function takes two observations as input and outputs their similarity.

SVM is a powerful algorithm, but also more compute-intensive than some of the less sophisticated classification algorithms. One of the advantages of SVM is that it works well on datasets that are not linearly separable.

Naïve Bayes

The Naïve Bayes algorithm uses Bayes theorem to train a classifier. The model trained by the Naïve Bayes algorithm is a probabilistic classifier. For a given observation, it calculates a probability distribution over a set of classes.

Bayes theorem describes the conditional or posterior probability of an event. The mathematical equation for Bayes theorem is shown next.

$$P(A \mid B) = \frac{P(B \mid A).P(A)}{P(B)}$$

In the preceding equation, A and B are events. P(A|B) is the posterior or conditional probability of A knowing that B has occurred. P(B|A) is the posterior probability of B given that A has occurred. P(A) and P(B) are the prior probabilities of A and B respectively.

The Naive Bayes algorithm assumes that all the features or predictor variables are independent. That is the reason it is called naïve. In theory, the Naive Bayes algorithm should be used only if the predictor variables are statistically independent; however, in practice, it works even when the independence assumption is not valid.

Naïve Bayes is particularly suited for high dimensional datasets. Although it is a simple algorithm, it often outperforms more sophisticated classification algorithms.

Decision Trees

The Decision Tree algorithm was covered under the section covering regression algorithms. As mentioned earlier, Decision Trees can be used for both regression and classification tasks. The algorithm works the same way in both cases except for one thing - the values stored at the terminal (leaf) nodes.

For regression tasks, each terminal node stores a numeric value; whereas for classification tasks, each terminal node stores a class label. Multiple leaves may have the same class label. To predict a label for an observation, a decision tree model starts at the root node of a decision tree and tests the features against the internal nodes until it arrives at a leaf node. The value at the leaf node is the predicted label.

Ensembles

The ensemble algorithms were covered under the section covering regression algorithms. Ensemble algorithms such as Random Forests and Gradient-Boosted Trees can also be used for classification tasks.

The logic for predicting a class label using Random Forests is different for classification tasks. A Random Forests model collects the predictions from its ensemble of decision trees and outputs the class label predicted by the maximum number of trees as its prediction.

Neural Network

Neural Network algorithms are inspired by biological neural networks. They try to mimic the brain. A commonly used neural network algorithm for classification tasks is the feedforward neural network. A classifier trained by the feedforward neural network algorithm is also known as a multi-layer perceptron classifier.

A multi-layer perceptron classifier consists of interconnected nodes (see Figure 8-7). A node is also referred to as a unit. The network of interconnected nodes is divided into multiple layers.

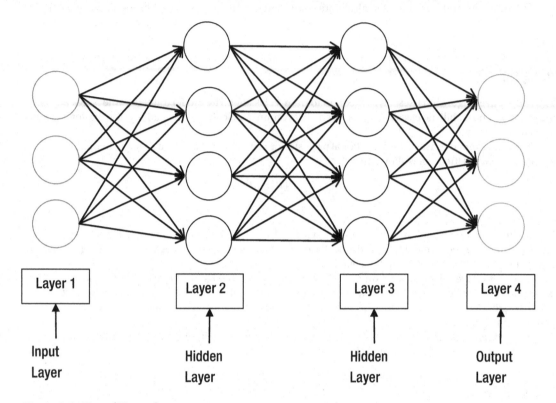

Figure 8-7. *Neural Network*

The first layer consists of the inputs to the classifier. It represents the features of an observation. Thus, the number of the nodes in the first layer is same as the number of input features.

The input layer is followed by one or more hidden layers. A neural network with two or more hidden layer is known as a deep neural network. Deep learning algorithms, which have recently become popular again, train models with multiple hidden layers.

A hidden layer can consist of any number of nodes or units. Generally, the predictive performance improves with more number of nodes in a hidden layer. Each node in a hidden layer accepts inputs from the all the nodes in the previous layer and produces an output value using an activation function.

The activation function generally used is the logistic (sigmoid) function. Thus, a single-layer feedforward neural network with no hidden layers is identical to a logistic regression model.

The last layer, also known as output units, represents label classes. The number of nodes in the output layer depends on the number of label classes. A binary classifier will have one node in the output layer. A k-class classifier will have k output nodes.

In a feedforward neural network, input data flows only forward from the input layer to the output layer through the hidden layers. There are no cycles.

The diagram in Figure 8-7 shows a feedforward neural network.

The feedforward neural network algorithm uses a technique known as backpropagation to train a model. During the training phase, prediction errors are fed back to the network. The algorithm uses this information for adjusting the weights of the edges connecting the nodes to minimize prediction errors. This process is repeated until the prediction errors converge to value less than a predefined threshold.

Generally, a neural network with one layer is sufficient in most cases. If more than one hidden layers are used, it is recommended to have the same number of nodes in each hidden layer.

Neural networks are better suited for classifying data that is not linearly separable. An example of a classification tasks involving non-linear data is shown in Figure 8-8.

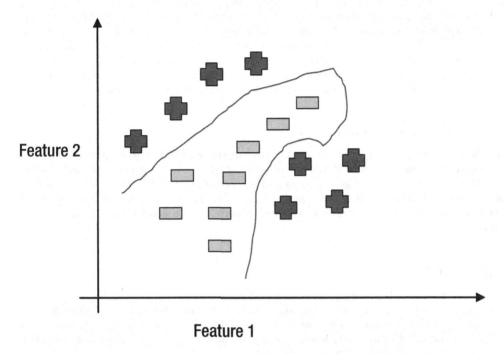

Figure 8-8. *Non-linear Classifier*

Neural networks have a few disadvantages. They are difficult to interpret. It is hard to explain what the nodes in the hidden layers represent. In addition, neural network algorithms are more compute intensive than simpler classification algorithms such as logistic regression.

Unsupervised Machine Learning Algorithms

An unsupervised machine learning algorithm is used when a dataset is unlabeled. It draws inferences from unlabeled datasets. Generally, the goal is to find hidden structure in unlabeled data. Unsupervised machine learning algorithms are generally used for clustering, anomaly detection, and dimensionality reduction.

The list of commonly used unsupervised machine learning algorithms includes k-means, Principal Component Analysis, and Singular Value Decomposition (SVD.

k-means

The k-means algorithm finds groupings or clusters in a dataset. It is an iterative algorithm that partitions data into k mutually exclusive clusters, where k is a number specified by a user.

The k-means algorithm uses a criterion known as *within-cluster sum-of-squares* for assigning observations to clusters. It iteratively finds cluster centers and assign observations to clusters such that within-cluster sum-of-squares is minimized.

The number of clusters in which data should be segmented is specified as an argument to the k-means algorithm.

Principal Components Analysis (PCA)

PCA is used for dimensionality reduction. It is a statistical method for reducing a large set of possibly correlated variables to a smaller set of uncorrelated variables, known as principal components. The number of principal components is less than or equal to the number of original variables.

The goal of PCA is to find the fewest number of variables responsible for the maximum amount of variability in a dataset. The first principal component is the variable with the largest variance. The second component is the variable with the second largest variance and it is statistically independent with respect to the first component. Similarly, the third component is the variable with the third largest variance and orthogonal to the first two components. This is true for each succeeding principal component. Thus, each principal component has the largest variance possible under the constraint that it is uncorrelated to the previous components.

Singular value decomposition (SVD)

Singular value decomposition (SVD) is one of the methods that can be used for PCA. It is a numerical method for finding the best approximation of a dataset using fewer dimensions. Using matrix factorization, it transforms a high-dimensional dataset into a dataset with fewer dimensions, while retaining important information. In other words, SVD identifies and orders statistically independent dimensions that account for the most variation in a dataset. Thus, it reduces a large set of correlated variables to a smaller set of uncorrelated variables.

Hyperparameter

As mentioned earlier, a machine learning algorithm fits a model over a dataset. In this process, it learns the model parameters. However, a machine learning algorithm also requires a few input parameters that determine the training time and predictive effectiveness of a trained model. These parameters, which are not directly learnt, but provided as inputs to a learning algorithm, are known as hyperparameters. Examples are shown later in the sections covering MLlib and Spark ML.

Model Evaluation

Before a model is deployed for use with new data, it is important to evaluate it on a test dataset. The predictive effectiveness or quality of a model can be evaluated using a few different metrics.

Generally, the evaluation metric depends on the machine learning task. Different metrics are used for linear regression, classification, clustering, and recommendation.

A simple model evaluation metric is accuracy. It is defined as the percentage of the labels correctly predicted by a model. For example, if a test dataset has 100 observations and a model correctly predicts the labels for 90 observations, its accuracy is 90%.

However, accuracy can be a misleading metric. For example, consider a tumors database, where each row has data about either a malignant or a benign tumor. In the context of machine learning, a malignant tumor is considered a positive sample and a benign tumor is considered a negative sample. Suppose we train a model that predicts whether a tumor is malignant (positive) or non-cancerous benign (negative). Is it a good model if it has 90% accuracy?

It depends on the test dataset. If the test dataset has 50% positive and 50% negative samples, our model is performing well. However, if the test dataset has only 1% positive and 99% negative samples, our model is worthless. We can generate a better model without using machine learning; a simple model that always classifies a sample as negative will have 99% accuracy. Thus, it has a better accuracy than our trained model, even though it incorrectly classifies all the positive samples.

The two commonly used metrics for evaluating a classifier or classification model are Area under Curve and F-measure.

Area Under Curve (AUC)

Area under Curve (AUC), also known as Area under ROC, is a metric generally used for evaluating binary classifiers (see Figure 8-9). It represents the proportion of the time a model correctly predicts the label for a random positive or negative observation. It can be graphically represented by plotting the rate of true positives predicted by a model against its rate of false positives. The best classifier has the largest area under the curve.

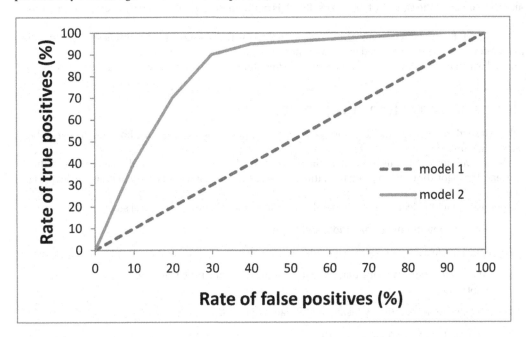

Figure 8-9. *Area under curve*

A model that just randomly guesses the label for an observation will have an AUC score of approximately 0.5. A model with an AUC score of 0.5 is considered worthless. A perfect model has an AUC score of 1.0. It predicts zero false positives and zero false negatives.

F-measure

F-measure, also known as F-score or F1 score, is another commonly used metric for evaluating classifiers. Let's define two other terms—recall and precision—before defining F-measure.

Recall is the fraction of the positive examples classified correctly by a model. The formula for calculating recall is shown next.

Recall = TP / (TP + FN), where TP = True Positives, and FN = False Negatives

Precision is the ratio of true positives to all positives classified by a model. It is calculated using the following formula.

Precision = TP / (TP + FP), where TP = True Positives, and FP = False Positives

The F-measure of a model is the harmonic mean of the recall and precision of a model. The formula for calculating the F-measure of a model is shown here.

F-measure = 2* (precision * recall) / (precision + recall)

The F-measure of a model takes a value between 0 and 1. The best model has an F-measure equal to 1, whereas a model with an F-score of 0 is the worst.

Root Mean Squared Error (RMSE)

The RMSE metric is generally used for evaluating models generated by regression algorithms. A related metric is mean squared error (MSE). An error in the context of a regression algorithm is the difference between the actual and predicted numerical label of an observation. As the name implies, MSE is the mean of the square of the errors. It is mathematically calculated by squaring the error for each observation and computing the mean of the square of the errors. RMSE is mathematically calculated by taking a square root of MSE.

RMSE and MSE represent training error. They indicate how well a model fits a training set. They capture the discrepancy between the observed labels and the labels predicted by a model.

A model with a lower MSE or RMSE represents a better fitting model than one with a higher MSE or RMSE.

Machine Learning High-level Steps

This is the last section before getting into the specifics of the Spark machine learning libraries. It covers the typical steps in a machine learning task.

The high-level steps generally depend on the type of a machine learning task and not so much on the machine learning algorithms. For a given task, the same steps can be used with different machine learning algorithms.

A supervised machine learning task generally consists of the following high-level steps.

1. Split data into training, validation, and test sets.

2. Select the features for training a model.

3. Fit a model on the training dataset using a supervised machine learning algorithm.

4. Tune the hyperparameters using the validation dataset.

5. Evaluate the model on a test dataset.

6. Apply the model to new data.

An unsupervised machine learning task generally consists of the following high-level steps.

1. Select the feature variables.

2. Fit a model using an unsupervised machine learning algorithm.

3. Evaluate the model using the right evaluation metrics.

4. Use the model.

Spark Machine Learning Libraries

Spark provides two machine learning libraries, MLlib and Spark ML (also known as the Pipelines API). These libraries enable high-performance machine learning on large datasets. Unlike machine learning libraries that can be used only with datasets that fit on a single machine, both MLlib and Spark ML are scalable. They make it possible to utilize a multi-node cluster for machine learning. In addition, since Spark allows an application to cache a dataset in memory, machine learning applications built with Spark ML or MLlib are fast.

MLlib is the first machine learning library that shipped with Spark. It is more mature than Spark ML. Both libraries provide higher-level abstractions for machine learning than the core Spark API (see Figure 8-10).

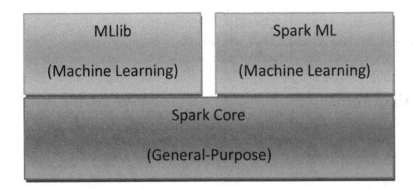

Figure 8-10. MLlib and Spark ML run on Spark

MLlib Overview

MLlib extends Spark for machine learning and statistical analysis. It provides a higher-level API than the Spark core API for machine learning and statistical analysis. It comes prepackaged with commonly used machine learning algorithms used for a variety of machine learning tasks. It also includes statistical utilities for different statistical analysis.

Integration with Other Spark Libraries

MLlib integrates with other Spark libraries such as Spark Streaming and Spark SQL (see Figure 8-11). It can be used with both batch and streaming data.

Figure 8-11. MLlib works with other Spark libraries

Data preparation steps such as data cleansing and feature engineering becomes easier with the DataFrame API provided by Spark SQL. Generally, the raw data cannot be used directly with machine learning algorithms. Features need to be extracted from the raw data.

Statistical Utilities

MLlib provides classes and functions for common statistical analysis. It supports summary statistics, correlations, stratified sampling, hypothesis testing, random data generation, and kernel density estimation.

Machine Learning Algorithms

MLlib can be used for common machine learning tasks such as regression, classification, clustering, anomaly detection, dimensionality reduction, and recommendation. The list of algorithms that come bundled with MLlib is ever growing. This section lists the algorithms shipped with MLlib at the time of writing this book.

Regression and Classification

- Linear regression

- Logistic regression

- Support Vector Machine

- Naïve Bayes

- Decision tree

- Random forest

- Gradient-boosted trees

- Isotonic regression

Clustering

- K-means

- Streaming k-means

- Gaussian mixture

- Power iteration clustering (PIC)

- Latent Dirichlet allocation (LDA)

Dimensionality Reduction

- Principal component analysis (PCA)

- Singular value decomposition (SVD)

Feature Extraction and Transformation

- TF-IDF

- Word2Vec

- Standard Scaler

- Normalizer

- Chi-Squared feature selection

- Elementwise product

Frequent pattern mining

- FP-growth
- Association rules
- PrefixSpan

Recommendation

- Collaborative filtering with Alternating Least Squares (ALS)

The MLlib API

The MLlib library can be used with applications developed in Scala, Java, Python, or R. This section covers the Scala version of the MLlib API. The classes and singleton objects provided by MLlib are available under the org.apache.spark.mllib package.

You can try the code examples shown in this section in the Spark REPL. Let's launch the Spark shell from a terminal.

```
path/to/SPARK_HOME/bin/spark-shell --master local[*]
```

Data Types

MLlib's primary data abstractions are Vector, LabeledPoint, and Rating. The machine learning algorithms and statistical utilities in MLlib operate on data represented by these abstractions.

Vector

The Vector type represents an indexed collection of Double-type values with zero-based index of type Int. It is generally used for representing the features of an observation in a dataset. Conceptually, a Vector of length n represents an observation with n features. In other words, it represents an element in an n-dimensional space.

The Vector type provided by MLlib should not be confused with the Vector type in the Scala collection library. They are different. The MLlib Vector type implements the concept of numeric vector from linear algebra. An application must import org.apache.spark.mllib.linalg.Vector to use the Vector trait provided by MLlib.

The MLlib library supports two types of vectors: dense and sparse. The MLlib Vector type is defined as a trait, so an application cannot directly create an instance of Vector. Instead, it should use the factory methods provided by MLlib to create either an instance of the DenseVector or SparseVector class. These two classes implement the Vector trait. The factory methods for creating an instance of the DenseVector or SparseVector class are defined in the Vectors object.

DenseVector

An instance of the DenseVector class stores a double-type value at each index position. It is backed by an array. A dense vector is generally used if a dataset does not have too many zero values. It can be created, as shown here.

```
import org.apache.spark.mllib.linalg._
val denseVector = Vectors.dense(1.0, 0.0, 3.0)
```

The dense method creates an instance of the DenseVector class from the values provided to it as arguments. A variant of the dense method takes an Array of Double type as an argument and returns an instance of the DenseVector class.

SparseVector

The SparseVector class represents a sparse vector, which stores only non-zero values. It is an efficient data type for storing a large dataset with many zero values. An instance of the SparseVector class is backed by two arrays; one stores the indices for non-zero values and the other stores the non-zero values.

A sparse vector can be created, as shown here.

```
import org.apache.spark.mllib.linalg._
val sparseVector = Vectors.sparse(10, Array(3, 6), Array(100.0, 200.0))
```

The sparse method returns an instance of the SparseVector class. The first argument to the sparse method is the length of a sparse vector. The second argument is an array specifying the indices for non-zero entries. The third argument is an array specifying the values. The indices and value array must have the same length.

Alternatively, a sparse vector can be created by specifying its length, and non-zero entries along with their positions, as shown here.

```
import org.apache.spark.mllib.linalg._
val sparseVector = Vectors.sparse(10, Seq((3, 100.0), (6, 200.0)))
```

LabeledPoint

The LabeledPoint type represents an observation in a labeled dataset. It contains both the label (dependent variable) and features (independent variables) of an observation. The label is stored as a Double-type value and the features are stored as a Vector type.

An RDD of LabeledPoints is MLlib's primary abstraction for representing a labeled dataset. Both regression and classification algorithms provided by MLlib operate only on RDD of LabeledPoints. Therefore, a dataset must be transformed to an RDD of LabeledPoints before it can be used to train a model.

Since the label field in a LabeledPoint is of type Double, it can represent both numerical and categorical labels. When used with a regression algorithm, the label in a LabeledPoint stores a numerical value. For binary classification, a label must be either 0 or 1. 0 represents a negative label and 1 represents a positive label. For multi-class classification, a label stores a zero-based class index of an observation.

LabeledPoints can be created, as shown here.

```
import org.apache.spark.mllib.linalg.Vectors
import org.apache.spark.mllib.regression.LabeledPoint

val positive = LabeledPoint(1.0, Vectors.dense(10.0, 30.0, 20.0))
val negative = LabeledPoint(0.0, Vectors.sparse(3, Array(0, 2), Array(200.0, 300.0)))
```

This code creates two LabeledPoints. The first LabeledPoint represents a positive observation with three features. The second LabeledPoint represents a negative observation with three features.

Rating

The Rating type is used with recommendation algorithms. It represents a user's rating for a product or item. A training dataset must be transformed to an RDD of Ratings before it can be used to train a recommendation model.

Rating is defined as a case class consisting of three fields. The first field is named user, which is of type Int. It represents a user identifier. The second field is named product, which is also of type Int. It represents a product or item identifier. The third field is named rating, which of type Double.

An instance of Rating can be created, as shown here.

```
import org.apache.spark.mllib.recommendation._

val rating = Rating(100, 10, 3.0)
```

This code creates an instance of the Rating class. This instance represents a rating of 3.0 given by a user with identifier 100 to a product with identifier 10.

Algorithms and Models

This section briefly describes MLlib's abstractions for representing machine learning algorithms and models.

A model in MLlib is represented by a class. MLlib provides different classes for representing models trained with different machine learning algorithms.

Similarly, a machine algorithm is represented by a class. MLlib also generally provides a companion singleton object with the same name for each machine learning algorithm class. It is more convenient to train a model using a singleton object representing a machine learning algorithm.

Training and using a model generally involves two key methods: train and predict. The train method is provided by the singleton objects representing machine learning algorithms. It trains a model with a given dataset and returns an instance of an algorithm-specific model class. The predict method is provided by the classes representing models. It returns a label for a given set of features.

Spark comes prepackaged with sample datasets that can be used to experiment with the MLlib API. For simplicity, the examples use those sample datasets.

Some of the data files are in the LIBSVM format. Each line stores an observation. The first column is a label. It is followed by the features, which are represented as offset:value, where offset is the index into the feature vector, and value is the value of a feature.

The MLlib library provides helper functions that create RDD[LabeledPoint] from files containing labeled data in the LIBSVM format. These methods are provided by the MLUtils object, which is available in the org.apache.spark.mllib.util package.

Regression Algorithms

The list of MLlib classes representing different regression algorithms includes LinearRegressionWithSGD, RidgeRegressionWithSGD, LassoWithSGD, ElasticNetRegression, IsotonicRegression, DecisionTree, GradientBoostedTrees, *and* RandomForest. MLlib also provides companion singleton objects with the same names. These classes and objects provide methods for training regression models.

This section describes the methods provided by the singleton objects for training a model.

train

The train method of a regression algorithm object trains or fits a linear regression model with a dataset provided to it as input. It takes an RDD of LabeledPoints as an argument and returns an algorithm-specific regression model. For example, the train method in the LinearRegressionWithSGD object returns an instance of the LinearRegressionModel class. Similarly, the train method in the DecisionTree object returns an instance of the DecisionTreeModel class.

The train method also takes a few additional algorithm-specific hyperparameters as arguments. For example, the train method in the RidgeRegressionWithSGD object takes as arguments the number of iterations of gradient descent to run, step size for each gradient descent iteration, regularization parameter, fraction of data to use in each iteration, and initial set of weights. Similarly, the train method in the GradientBoostedTrees object takes a boosting strategy as an argument.

The following example demonstrates how to train a model with LinearRegressionWithSGD. You can use the other regression algorithms in a similar way.

```
import org.apache.spark.mllib.linalg.Vectors
import org.apache.spark.mllib.regression.LabeledPoint
import org.apache.spark.mllib.regression.LinearRegressionWithSGD

// create a RDD from a text file
val lines = sc.textFile("data/mllib/ridge-data/lpsa.data")
```

```
lines: org.apache.spark.rdd.RDD[String] = MapPartitionsRDD[1] at textFile at <console>:24
```

```
// transform the raw dataset into a RDD of LabelelPoint
val labeledPoints = lines.map { line =>
                // split each line into a label and features
                val Array(rawLabel, rawFeatures) = line.split(',')
                // extract individual features and convert them to type Double
                val features = rawFeatures.split(' ').map(_.toDouble)
                // create a LabeledPoint for each input line
                LabeledPoint(rawLabel.toDouble, Vectors.dense(features))
            }
```

```
labeledPoints: org.apache.spark.rdd.RDD[org.apache.spark.mllib.regression.LabeledPoint] =
MapPartitionsRDD[2] at map at <console>:26
```

```
// cache the dataset since the training step is iterative.
labeledPoints.cache

val numIterations = 100

// train a model
val lrModel = LinearRegressionWithSGD.train(labeledPoints, numIterations)
```

```
lrModel: org.apache.spark.mllib.regression.LinearRegressionModel = org.apache.spark.mllib.
regression.LinearRegressionModel: intercept = 0.0, numFeatures = 8
```

```
// check the model parameters
val intercept = lrModel.intercept
```

```
intercept: Double = 0.0
```

```
val weights = lrModel.weights
```

```
weights: org.apache.spark.mllib.linalg.Vector = [0.5808575763272221,0.1893000148294698,
0.2803086929991066,0.11108341817778758,0.4010473965597894,-0.5603061626684255,
-0.5804740464000983,0.8742741176970946]
```

This model will be used in the section that covers regression models.

trainRegressor

The trainRegressor method is provided by the singleton objects representing tree-based algorithms such as DecisionTree and RandomForest. It trains or fits a non-linear regression model with a dataset provided to it as input. It takes an RDD of LabeledPoint and algorithm-specific hyperparameters as arguments. For example, in case of RandomForest, it takes as arguments the number of trees, maximum depth of a tree, maximum number of bins, random seed for bootstrapping, categorical features, and number of features to consider for splits at each node.

The trainRegressor method returns an algorithm-specific non-linear regression model. For example, the trainRegressor method in the DecisionTree object returns an instance of the DecisionTreeModel class. Similarly, the trainRegressor method in the RandomForest object returns an instance of the RandomForestModel class.

The following example demonstrates how to train a model with RandomForest. We will use the same dataset that we used earlier to train a model with LinearRegressionWithSGD.

```
import org.apache.spark.mllib.linalg.Vectors
import org.apache.spark.mllib.regression.LabeledPoint
import org.apache.spark.mllib.tree.RandomForest

val lines = sc.textFile("data/mllib/ridge-data/lpsa.data")

// transform the raw dataset into a RDD of LabelelPoint
val labeledPoints = lines.map { line =>
                    // split each line into a label and features
                    val Array(rawLabel, rawFeatures) = line.split(',')
                    // extract individual features and convert them to type Double
                    val features = rawFeatures.split(' ').map(_.toDouble)
                    // create a LabeledPoint for each input line
                    LabeledPoint(rawLabel.toDouble, Vectors.dense(features))
                    }

// cache the dataset since the training step is iterative.
labeledPoints.cache
```

```
// Initialize the hyperparameters for the RandomForest algorithm

/*
categoricalFeaturesInfo input is a Map storing arity of categorical features.
An Map entry (n -> k) indicates that feature n is categorical with k categories indexed
from 0: {0, 1, ..., k-1}.
*/
val categoricalFeaturesInfo = Map[Int, Int]()  // all features are continuous.

// numTrees specifies number of trees in the random forest.
val numTrees = 3 // Use more in practice.

/*
featureSubsetStrategy specifies number of features to consider for splits at each node.
MLlib supports: "auto", "all", "sqrt", "log2", "onethird".
If "auto" is specified, the algorithm choses a value based on numTrees: if numTrees == 1,
featureSubsetStrategy is set to "all"; other it is set to "onethird".
*/
val featureSubsetStrategy = "auto"

/*
impurity specifies the criterion used for information gain calculation.
Supported values: "variance"
*/
val impurity = "variance"

/*
maxDepth specifies the maximum depth of the tree. Depth 0 means 1 leaf node; depth 1 means 1
internal node + 2 leaf nodes.
Suggested value: 4
*/
val maxDepth = 4

/*
maxBins specifies the maximum number of bins to use for splitting features.
Suggested value: 100
*/
val maxBins = 32

// Train a model.
val rfModel = RandomForest.trainRegressor(labeledPoints, categoricalFeaturesInfo,
  numTrees, featureSubsetStrategy, impurity, maxDepth, maxBins)
```

```
rfModel: org.apache.spark.mllib.tree.model.RandomForestModel =
TreeEnsembleModel regressor with 3 trees
```

Regression Models

The classes representing regression models include LinearRegressionModel, RidgeRegressionModel, LassoModel, IsotonicRegressionModel, DecisionTreeModel, GradientBoostedTreesModel, and RandomForestModel. Instances of these classes are returned by the train or trainRegressor methods of the objects mentioned in the previous section.

The commonly used methods in these classes are briefly described next.

predict

The predict method of a regression model returns a numerical label for a given set of features. It takes a Vector as an argument and returns a value of type Double.

A variant of the predict method takes an RDD of Vector as argument and returns an RDD of Double. Thus, it can be used to either predict a label for an observation or a dataset.

The following code snippet uses an instance of the LinearRegressionModel class trained earlier. It calculates the mean squared error for the model.

```
// get actual and predicted label for each observation in the training set
val observedAndPredictedLabels = labeledPoints.map { observation =>
  val predictedLabel = lrModel.predict(observation.features)
  (observation.label, predictedLabel)
}
```

```
observedAndPredictedLabels: org.apache.spark.rdd.RDD[(Double, Double)] =
MapPartitionsRDD[161] at map at <console>:32
```

```
// calculate square of difference between predicted and actual label for each observation
val squaredErrors = observedAndPredictedLabels.map{case(actual, predicted) =>
                      math.pow((actual - predicted), 2)
                    }
```

```
squaredErrors: org.apache.spark.rdd.RDD[Double] = MapPartitionsRDD[162] at map at
<console>:34
```

```
// calculate the mean of squared errors.
val meanSquaredError = squaredErrors.mean()
```

```
meanSquaredError: Double = 6.207597210613578
```

An instance of the LinearRegressionModel class was used in the preceding example. You can similarly use instances of DecisionTreeModel, RandomForestModel, and other model classes.

save

The save method persists a trained model to disk. It takes a SparkContext and path as arguments and saves the source model to the given path. The saved model can be later read with the load method.

```
lrModel.save(sc, "models/lr-model")
```

load

The load method is defined in the companion model objects. It generates an instance of a model from a previously saved model. It takes as arguments a SparkContext and the path to a saved model and returns an instance of a model class.

```
import org.apache.spark.mllib.regression.LinearRegressionModel
val savedLRModel = LinearRegressionModel.load(sc, "models/lr-model")
```

```
savedLRModel: org.apache.spark.mllib.regression.LinearRegressionModel = org.apache.spark.
mllib.regression.LinearRegressionModel: intercept = 0.0, numFeatures = 8
```

```
// check the model parameters
val intercept = savedLRModel.intercept
```

```
intercept: Double = 0.0
```

```
val weights = savedLRModel.weights
```

```
weights: org.apache.spark.mllib.linalg.Vector = [0.5808575763272221,0.1893000148294698,
0.2803086929991066,0.11108341817778758,0.4010473965597894,-0.5603061626684255,
-0.5804740464000983,0.8742741176970946]
```

toPMML

The toPMML method exports a model in Predictive Model Markup Language (PMML) format. PMML is an XML-based format for describing and serializing models generated by machine learning algorithms. It enables different applications to share models. With PMML, you can train a model in one application and use it from another application.

Multiple variants of the toPMML method are available. These variants allow you to export a model to a string, OutputStream, local file, or a distributed file system.

The following example shows how to export a model in PMML format to a String.

```
val lrModelPMML = lrModel.toPMML
```

```
lrModelPMML: String =
"<?xml version="1.0" encoding="UTF-8" standalone="yes"?>
<PMML xmlns="http://www.dmg.org/PMML-4_2">
    <Header description="linear regression">
        <Application name="Apache Spark MLlib" version="1.5.2"/>
        <Timestamp>2015-11-21T09:51:52</Timestamp>
    </Header>
    <DataDictionary numberOfFields="9">
        <DataField name="field_0" optype="continuous" dataType="double"/>
        <DataField name="field_1" optype="continuous" dataType="double"/>
```

```
<DataField name="field_2" optype="continuous" dataType="double"/>
<DataField name="field_3" optype="continuous" dataType="double"/>
<DataField name="field_4" optype="continuous" dataType="double"/>
<DataField name="field_5" optype="continuous" dataType="double"/>
...
```

Classification Algorithms

The list of MLlib classes representing different classification algorithms includes LogisticRegressionWithSGD, LogisticRegressionWithLBFGS, SVMWithSGD, NaiveBayes, DecisionTree, GradientBoostedTrees, and RandomForest. MLlib also provides companion singleton objects with the same names. These classes and objects provide methods for training classification models, which are also referred to as classifiers.

This section briefly describes the methods provided by the singleton objects representing classification algorithms for training a model.

train

The train method of a classification object trains or fits a classification model with a dataset provided to it as input. It takes an RDD of LabeledPoint as an argument and returns an algorithm-specific classification model. For example, the train method in the LogisticRegressionWithSGD object returns an instance of the LogisticRegressionModel class. Similarly, the train method in the NaiveBayes object returns an instance of the NaiveBayesModel class.

In addition to an RDD of LabeledPoint, the train method also takes algorithm-specific hyperparameters as arguments. For example, the train method in the NaiveBayes object takes the smoothing parameter as an argument. Similarly, the *train* method in the SVMWithSGD object takes the number of iterations of gradient descent to run, step size for each gradient descent iteration, regularization parameter, fraction of data to use in each iteration, and initial set of weights as arguments.

The following example trains a model with SVMWithSGD. You can use singleton objects representing other classification algorithms in a similar way.

```
import org.apache.spark.mllib.linalg.Vectors
import org.apache.spark.mllib.regression.LabeledPoint
import org.apache.spark.mllib.classification.SVMWithSGD
import org.apache.spark.mllib.util.MLUtils

// Load binary labeled data from a file in LIBSVM format
val labeledPoints = MLUtils.loadLibSVMFile(sc, "data/mllib/sample_libsvm_data.txt")
```

```
labeledPoints: org.apache.spark.rdd.RDD[org.apache.spark.mllib.regression.LabeledPoint] =
MapPartitionsRDD[6] at map at MLUtils.scala:112
```

```
// Split data into training (60%), validation (20%) and test (20%) set.
val Array(trainingData, validationData, testData) = labeledPoints.randomSplit(Array(0.6,
0.2, 0.2))
```

```
trainingData: org.apache.spark.rdd.RDD[org.apache.spark.mllib.regression.LabeledPoint] =
MapPartitionsRDD[7] at randomSplit at <console>:27
validationData: org.apache.spark.rdd.RDD[org.apache.spark.mllib.regression.LabeledPoint] =
MapPartitionsRDD[8] at randomSplit at <console>:27
testData: org.apache.spark.rdd.RDD[org.apache.spark.mllib.regression.LabeledPoint] =
MapPartitionsRDD[9] at randomSplit at <console>:27
```

```
// Persist the training data in memory to speed up the training step
trainingData.cache()
```

```
val numIterations = 100
```

```
// Fit a SVM model on the training dataset
val svmModel = SVMWithSGD.train(trainingData, numIterations)
```

```
svmModel: org.apache.spark.mllib.classification.SVMModel = org.apache.spark.mllib.
classification.SVMModel: intercept = 0.0, numFeatures = 692, numClasses = 2,
threshold = 0.0
```

trainClassifier

The trainClassifier method is provided by the singleton objects representing tree-based algorithms such as DecisionTree and RandomForest. It fits a non-linear model or classifier with a dataset provided to it as input. Similar to trainRegressor, it takes an RDD of LabeledPoints and algorithm-specific hyperparameters as arguments. It returns an algorithm-specific model.

Classification Models

The classes representing classification models include LogisticRegressionModel, NaiveBayesModel, SVMModel, DecisionTreeModel, GradientBoostedTreesModel, *and* RandomForestModel. Instances of these classes are returned by the train or trainClassifier methods of the corresponding classification algorithms. The commonly used methods from these classes are briefly described next.

predict

The predict method returns a class or categorical label for a given set of features. It takes a Vector as an argument and returns a value of type Double.

The following example uses an instance of the SVMModel class trained earlier. It generates a predicted label for each observation in the test dataset. We will use this later to evaluate our model.

```
// get actual and predicted label for each observation in the test set
val predictedAndActualLabels   = testData.map { observation =>
  val predictedLabel = svmModel.predict(observation.features)
  (predictedLabel, observation.label)
}
```

```
predictedAndActualLabels: org.apache.spark.rdd.RDD[(Double, Double)] = MapPartitionsRDD[213]
```

```
val fivePredAndActLabel = predictedAndActualLabels.take(5)
```

```
fivePredAndActLabel: Array[(Double, Double)] = Array((1.0,1.0), (0.0,0.0), (1.0,1.0),
(1.0,1.0), (1.0,1.0))
```

Note that the pairs that you get for fivePredAndActLabel may be different from shown in the preceding example. The reason is that the input dataset was randomly split into training, validation, and test data. Each time randomSplit is called, it may return different observations in testData. The important thing to look for is whether the predicted labels are same as the observed or actual labels.

A variant of the predict method takes an RDD[Vector] as an argument and returns an RDD[Double].

```
val predictedLabels = svmModel.predict(testData.map(_.features))
```

```
predictedLabels: org.apache.spark.rdd.RDD[Double] = MapPartitionsRDD[215] at mapPartitions
at GeneralizedLinearAlgorithm.scala:69
```

```
val fivePredictedLabels = predictedLabels.take(5)
```

```
fivePredictedLabels: Array[Double] = Array(1.0, 0.0, 1.0, 1.0, 1.0)
```

As explained earlier, the results that you get may be different from shown in the preceding example.

save

The save method persists a trained model to disk. It takes a SparkContext and path as arguments and saves the source model to the given path. The saved model can be later read with the load method.

```
svmModel.save(sc, "models/svm-model")
```

load

The load method is defined in the companion model objects. It generates an instance of a model from a previously saved model. It takes a SparkContext and the path to a saved model as arguments and returns an instance of a model.

```
import org.apache.spark.mllib.classification.SVMModel
val savedSVMModel = SVMModel.load(sc, "models/svm-model")
```

toPMML

The toPMML method exports a trained model in PMML format. Multiple variants are available to export a model to a string, OutputStream, local file or a distributed file system.

```
val svmModelPMML = model.toPMML
```

```
svmModelPMML: String =
"<?xml version="1.0" encoding="UTF-8" standalone="yes"?>
<PMML xmlns="http://www.dmg.org/PMML-4_2">
    <Header description="linear SVM">
        <Application name="Apache Spark MLlib" version="1.5.2"/>
        <Timestamp>2015-11-21T17:45:09</Timestamp>
    </Header>
    <DataDictionary numberOfFields="693">
        <DataField name="field_0" optype="continuous" dataType="double"/>
        <DataField name="field_1" optype="continuous" dataType="double"/>
        <DataField name="field_2" optype="continuous" dataType="double"/>
        <DataField name="field_3" optype="continuous" dataType="double"/>
        <DataField name="field_4" optype="continuous" dataType="double"/>
        <DataField name="field_5" optype="continuous" dataType="double"/>
        ...
```

Clustering Algorithms

The list of MLlib classes representing different clustering algorithms includes KMeans, StreamingKMeans, GaussianMixture, LDA, and PowerIterationClustering. MLlib also provides companion singleton objects with the same names.

The methods provided for training clustering models are briefly described next.

train

The train method is provided by the clustering-related singleton objects. It takes an RDD of Vector and algorithm-specific hyperparameters as arguments and returns an algorithm-specific clustering model.

The hyperparameter arguments and the type of the returned model depend on the clustering algorithm. For example, the hyperparameters accepted by the train method in the KMeans object include the number of clusters, maximum number of iterations in each run, number of parallel runs, initialization modes, and random seed value for cluster initialization. It returns an instance of the KMeansModel class.

The following example code uses the KMeans object to train a KMeansModel.

```scala
import org.apache.spark.mllib.clustering.KMeans
import org.apache.spark.mllib.linalg.Vectors

// Load data
val lines = sc.textFile("data/mllib/kmeans_data.txt")

// Convert each text line into an array of Double
val arraysOfDoubles = lines.map{line => line.split(' ').map(_.toDouble)}

// Transform the parsed data into a RDD[Vector]
val vectors = arraysOfDoubles.map{a => Vectors.dense(a)}.cache()

val numClusters = 2
val numIterations = 20

// Train a KMeansModel
val kMeansModel = KMeans.train(vectors, numClusters, numIterations)
```

run

The run method is provided by the clustering-related classes. Similar to the train method, it takes an RDD of Vector as argument and returns an algorithm-specific model. For the example, the *run* method in the PowerIterationClustering class returns an instance of the PowerIterationClusteringModel class.

The following example code uses an instance of the KMeans class to train a KMeansModel.

```scala
import org.apache.spark.mllib.clustering.KMeans
import org.apache.spark.mllib.linalg.Vectors

// Load data
val lines = sc.textFile("data/mllib/kmeans_data.txt")

// Convert each text line into an array of Double
val arraysOfDoubles = lines.map{line => line.split(' ').map(_.toDouble)}

// Transform the parsed data into a RDD[Vector]
val vectors = arraysOfDoubles.map{a => Vectors.dense(a)}.cache()

val numClusters = 2
val numIterations = 20

// Create an instance of the KMeans class and set the hyperparameters
val kMeans = new KMeans().setMaxIterations(numIterations).setK(numClusters)

// Train a KMeansModel
val kMeansModel = kMeans.run(vectors)
```

Clustering Models

The classes representing clustering models include KMeansModel, GaussianMixtureModel, PowerIterationClusteringModel, StreamingKMeansModel, *and* DistributedLDAModel. Instances of these classes are returned by the train methods of the clustering algorithm related objects and run methods of the clustering algorithm related classes.

The commonly used methods from the KMeansModel class are briefly described next.

predict

The predict method returns a cluster index for a given observation. It takes a Vector as an argument and returns a value of type Int.

The following example code determines the cluster indices for a couple of observation using the KMeansModel trained earlier.
```
val obs1 = Vectors.dense(0.0, 0.0, 0.0)
val obs2 = Vectors.dense(9.0, 9.0, 9.0)

val clusterIndex1 = kMeansModel.predict(obs1)
```

```
clusterIndex1: Int = 1
```

```
val clusterIndex2 = kMeansModel.predict(obs2)
```

```
clusterIndex2: Int = 0
```

A variant of the predict method maps a list of observations to their cluster indices. It takes an RDD of Vector as argument and returns an RDD of Int.

The following example code calculates the cluster indices for all the training data that we used to train a KMeansModel earlier.

```
val clusterIndicesRDD = kMeansModel.predict(vectors)
val clusterIndices = clusterIndicesRDD.collect()
```

```
clusterIndices: Array[Int] = Array(1, 1, 1, 0, 0, 0)
```

computeCost

The computeCost method returns the sum of the squared distances of the observations from their nearest cluster centers. It can be used to evaluate a KMeans model.
```
// Compute Within Set Sum of Squared Errors
val WSSSE = kMeansModel.computeCost(vectors)
```

```
WSSSE: Double = 0.11999999999994547
```

save

The save method persists a trained clustering model to disk. It takes a SparkContext and path as arguments and saves the source model to the given path. The saved model can be later read with the load method.

```
kMeansModel.save(sc, "models/kmean")
```

load

The load method is defined in the companion model objects. It generates an instance of a model from a previously saved model. It takes a SparkContext and the path to a saved model and returns an instance of a clustering model.

```
import org.apache.spark.mllib.clustering.KMeansModel
val savedKMeansModel = KMeansModel.load(sc, "models/kmean")
```

toPMML

The toPMML method exports a trained clustering model in PMML format. Multiple variants are available to export a model to a string, OutputStream, local file or a distributed file system.

```
val kMeansModelPMML = kMeansModel.toPMML()
```

```
kMeansModelPMML: String =
"<?xml version="1.0" encoding="UTF-8" standalone="yes"?>
<PMML xmlns="http://www.dmg.org/PMML-4_2">
    <Header description="k-means clustering">
        <Application name="Apache Spark MLlib" version="1.5.2"/>
        <Timestamp>2015-11-22T08:14:43</Timestamp>
    </Header>
    <DataDictionary numberOfFields="3">
        <DataField name="field_0" optype="continuous" dataType="double"/>
        <DataField name="field_1" optype="continuous" dataType="double"/>
        <DataField name="field_2" optype="continuous" dataType="double"/>
    </DataDictionary>
    <ClusteringModel modelName="k-means" functionName="clustering" modelClass="centerBased"
    numberOfClusters="2">
        <MiningSchema>
            <MiningField name="field_0" usageType="active"/>
            ...
```

Recommendation Algorithms

MLlib supports collaborative filtering, which learns latent factors describing users and products from a dataset containing only user identifiers, product identifiers, and ratings. Collaborative filtering-based recommendation system can be developed in MLlib using the ALS (alternating least squares) algorithm. MLlib provides a class named ALS, which implements Alternating Least Squares matrix factorization. It also provides a companion singleton object with the same name.

MLlib supports both ratings and implicit feedbacks. A rating is available when a user explicitly rates a product. For example, users rate movies and shows on Netflix. Similarly, users rate songs on iTunes, Spotify, Pandora, and other music services. However, sometimes an explicit rating is not available, but an implicit preference can be determined from user activities. For example, purchase of a product by a user conveys user's implicitly feedback for a product. Similarly, a user provides implicit feedback for a product by clicking a like or share button.

The methods provided by the ALS object for training a recommendation model are briefly described next.

train

The train method of the ALS object trains or fits a MatrixFactorizationModel model with an RDD of Rating. It takes an RDD of Rating and ALS-specific hyperparameters as arguments and returns an instance of the MatrixFactorizationModel class. The hyperparameters for ALS include the number of latent features, number of iterations, regularization factor, level of parallelism, and random seed. The last three are optional.

The following example code uses explicit ratings and the ALS object to train a recommendation model.

```
import org.apache.spark.mllib.recommendation.ALS
import org.apache.spark.mllib.recommendation.Rating

// Create a RDD[String] from a dataset
val lines = sc.textFile("data/mllib/als/test.data")

// Transform each text line into a Rating
val ratings = lines.map {line =>
                    val Array(user, item, rate) = line.split(',')
                    Rating(user.toInt, item.toInt, rate.toDouble)
            }

val rank = 10
val numIterations = 10

// Train a MatrixFactorizationModel
val mfModel = ALS.train(ratings, rank, numIterations, 0.01)
```

trainImplicit

The trainImplicit method can be used when only implicit user feedback for a product is available. Similar to the train method, it takes an RDD of Rating and ALS-specific hyperparameters as arguments and returns an instance of the MatrixFactorizationModel class.

The following code snippet uses implicit feedbacks and the ALS object to train a recommendation model.

```
val rank = 10
val numIterations = 10
val alpha = 0.01
val lambda = 0.01

// Train a MatrixFactorizationModel
val mfModel = ALS.trainImplicit(feedback, rank, numIterations, lambda, alpha)
```

Recommendation Models

A recommendation model is represented by an instance of the `MatrixFactorizationModel` class. MLlib also provides a companion object with the same name. The commonly used methods are briefly described next.

predict

The `predict` method in the `MatrixFactorizationModel` class returns a rating for a given user and product. It takes a user id and product id, which are of type `Int`, as arguments and returns a rating, which is of type `Double`.

```
val userId = 1
val prodId = 1

val predictedRating = mfModel.predict(userId, prodId)
```

```
predictedRating: Double = 4.997277769473439
```

A variant of the `predict` method accepts an RDD of pairs of user id and product id and returns an RDD of Rating. Thus, it can be used to predict both the rating for a single user-product pair or a list of user-product pairs.

```
// Create a RDD of user-product pairs
val usersProducts = ratings.map { case Rating(user, product, rate) => (user, product) }

// Generate predictions for all the user-product pairs
val predictions = mfModel.predict(usersProducts)

// Check the first five predictions
val firstFivePredictions = predictions.take(5)
```

```
firstFivePredictions: Array[org.apache.spark.mllib.recommendation.Rating] = Array(Rating
(4,4,4.9959530008728334), Rating(4,1,1.000928540744371), Rating(4,2,4.9959530008728334),
Rating(4,3,1.000928540744371), Rating(1,4,1.0006636186423625))
```

recommendProducts

The `recommendProducts` method recommends the specified number of products for a given user. It takes a user id and number of products to recommend as arguments and returns an Array of Rating. Each Rating object includes the given user id, product id and a predicted rating score. The returned Array is sorted by rating score in descending order. A high rating score indicates a strong recommendation.

```
val userId = 1
val numProducts = 3

val recommendedProducts = mfModel.recommendProducts(userId, numProducts)
```

```
recommendedProducts: Array[org.apache.spark.mllib.recommendation.Rating] = Array(Rating
(1,1,4.997277769473439), Rating(1,3,4.997277769473439), Rating(1,4,1.0006636186423625)).
```

recommendProductsForUsers

The recommendProductsForUsers method recommends the specified number of top products for all users. It takes the number of products to recommend as an argument and returns an RDD of users and corresponding top recommended products.

```
val numProducts = 2
val recommendedProductsForAllUsers = mfModel.recommendProductsForUsers(numProducts)
```

```
val rpFor4Users = recommendedProductsForAllUsers.take(4)
rpFor4Users: Array[(Int, Array[org.apache.spark.mllib.recommendation.Rating])] =
Array((1,Array(Rating(1,1,4.997277769473439), Rating(1,3,4.997277769473439))),
(2,Array(Rating(2,1,4.997277769473439), Rating(2,3,4.997277769473439))),
(3,Array(Rating(3,4,4.9959530008728334), Rating(3,2,4.9959530008728334))),
(4,Array(Rating(4,4,4.9959530008728334), Rating(4,2,4.9959530008728334))))
```

recommendUsers

The recommendUsers method recommends the specified number of users for a given product. This method returns a list of users who are most likely to be interested in a given product. It takes a product id and number of users to recommend as arguments and returns an Array of Rating. Each Rating object includes a user id, the given product id and a score in the rating field. The array is sorted by rating score in descending order...

```
val productId = 2
val numUsers = 3
```

```
val recommendedUsers = mfModel.recommendUsers(productId, numUsers)
```

```
recommendedUsers: Array[org.apache.spark.mllib.recommendation.Rating] =
Array(Rating(4,2,4.9959530008728334), Rating(3,2,4.9959530008728334),
Rating(1,2,1.0006636186423625))
```

recommendUsersForProducts

The recommendUsersForProducts method recommends the specified number of users for all products. It takes the number of users to recommend as an argument and returns an RDD of products and corresponding top recommended users.

```
val numUsers = 2
val recommendedUsersForAllProducts = mfModel.recommendUsersForProducts(numUsers)
val ruFor4Products = recommendedUsersForAllProducts.take(4)
```

```
ruFor4Products: Array[(Int, Array[org.apache.spark.mllib.recommendation.Rating])] =
Array((1,Array(Rating(1,1,4.997277769473439), Rating(2,1,4.997277769473439))),
(2,Array(Rating(4,2,4.9959530008728334), Rating(3,2,4.9959530008728334))),
(3,Array(Rating(1,3,4.997277769473439), Rating(2,3,4.997277769473439))),
(4,Array(Rating(4,4,4.9959530008728334), Rating(3,4,4.9959530008728334))))
```

save

The save method persists a MatrixFactorizationModel to disk. It takes a SparkContext and path as arguments and saves the source model to the given path. The saved model can be later read with the load method.

```
mfModel.save(sc, "models/mf-model")
```

load

The load method, defined in the") MatrixFactorizationModel object, can be used to read a previously saved model from a file. It takes a SparkContext and the path to a saved model and returns an instance of the MatrixFactorizationModel class.

```
import org.apache.spark.mllib.recommendation.MatrixFactorizationModel
val savedMfModel = MatrixFactorizationModel.load(sc, "models/mf-model")
```

Model Evaluation

As previously mentioned, evaluating a machine learning model before it is used with new data is an important step. We also earlier discussed some of the quantitative metrics that are commonly used for evaluating the effectiveness of a model. Fortunately, we do not need to manually compute these metrics.

MLlib comes prepackaged with classes that make it easy to evaluate models. These classes are available in the org.apache.spark.mllib.evaluation package. The list of model evaluation related classes includes BinaryClassificationMetrics, MulticlassMetrics, MultilabelMetrics, RankingMetrics, and RegressionMetrics.

Regression Metrics

The RegressionMetrics class can be used for evaluating models generated by regression algorithms. It provides methods for calculating mean squared error, root mean squared error, mean absolute error, R2, and other metrics.

The following example demonstrates how to use an instance of the RegressionMetrics class to evaluate a model trained with a regression algorithm.

```
import org.apache.spark.mllib.linalg.Vectors
import org.apache.spark.mllib.regression.LabeledPoint
import org.apache.spark.mllib.regression.LinearRegressionWithSGD
import org.apache.spark.mllib.evaluation.RegressionMetrics

// create a RDD from a text file
val lines = sc.textFile("data/mllib/ridge-data/lpsa.data")

// Transform the raw dataset into a RDD of LabelelPoint
val labeledPoints = lines.map { line =>
                    // split each line into a label and features
                    val Array(rawLabel, rawFeatures) = line.split(',')
                    // extract individual features and convert them to type Double
                    val features = rawFeatures.split(' ').map(_.toDouble)
```

```scala
                    // create a LabeledPoint for each input line
                    LabeledPoint(rawLabel.toDouble, Vectors.dense(features))
          }

// Cache the dataset
labeledPoints.cache

val numIterations = 100

// Train a model
val lrModel = LinearRegressionWithSGD.train(labeledPoints, numIterations)

// Create a RDD of actual and predicted labels
val observedAndPredictedLabels = labeledPoints.map { observation =>
  val predictedLabel = lrModel.predict(observation.features)
  (observation.label, predictedLabel)
}

// Create an instance of the RegressionMetrics class
val regressionMetrics = new RegressionMetrics(observedAndPredictedLabels)

// Check the various evaluation metrics
val mse = regressionMetrics.meanSquaredError
```

```scala
mse: Double = 6.207597210613578
```

```scala
val rmse = regressionMetrics.rootMeanSquaredError
```

```scala
rmse: Double = 2.491505009148803
```

```scala
val mae = regressionMetrics.meanAbsoluteError
```

```scala
mae: Double = 2.3439822940073354
```

Binary Classification Metrics

The BinaryClassificationMetrics class can be used for evaluating binary classifiers. It provides methods for calculating receiver operating characteristic (ROC) curve, area under the receiver operating characteristic (AUC) curve, and other metrics.

The following example demonstrates how to use an instance of the BinaryClassificationMetrics class to evaluate a binary classifier.

```scala
import org.apache.spark.mllib.linalg.Vectors
import org.apache.spark.mllib.regression.LabeledPoint
import org.apache.spark.mllib.classification.SVMWithSGD
import org.apache.spark.mllib.util.MLUtils
import org.apache.spark.mllib.evaluation.BinaryClassificationMetrics
```

```
// Load binary labeled data from a file in LIBSVM format
val labeledPoints = MLUtils.loadLibSVMFile(sc, "data/mllib/sample_libsvm_data.txt")

// Split data into training (80%) and test (20%) set.
val Array(trainingData, testData) = labeledPoints.randomSplit(Array(0.8, 0.2))

// Persist the training data in memory to speed up the training step
trainingData.cache()

val numIterations = 100

// Fit a SVM model on the training dataset
val svmModel = SVMWithSGD.train(trainingData, numIterations)

// Clear the prediction threshold so that the model returns probabilities
svmModel.clearThreshold

// get actual and predicted label for each observation in the test set
val predictedAndActualLabels   = testData.map { observation =>
  val predictedLabel = svmModel.predict(observation.features)
  (predictedLabel, observation.label)
}

// Create an instance of the BinaryClassificationMetrics class
val metrics = new BinaryClassificationMetrics(predictedAndActualLabels)

// Get area under curve
val auROC = metrics.areaUnderROC()
```

```
auROC: Double = 1.0
```

Multiclass Classification Metrics

The MulticlassMetrics class can be used for evaluating multi-class or multi-nominal classifiers. In a multi-class classification task, a label is not binary. An observation can take one of many possible labels. For example, a model that recognizes the images of animals is a multi-class classifier. An image can have the label *cat, dog, lion, elephant,* or some other label.

For evaluating a multi-nominal classifier, the MulticlassMetrics class provides methods for calculating precision, recall, F-measure, and other metrics.

We will show an example that uses the MulticlassMetrics class later in this chapter.

Multilabel Classification Metrics

The MultilabelMetrics class can be used for evaluating multi-label classifiers. In a multi-label classification task, an observation can have more than one label. The difference between a between a multi-label and multi-class dataset is that labels are not mutually exclusive in a multi-label classification task, whereas labels are mutually exclusive in a multi-class classification task. An observation in a multi-class classification task can take on only one of many labels.

An example of a multi-label classifier is a model that classifies an animal into different categories such as mammal, reptile, fish, bird, aquatic, terrestrial, or amphibian. An animal can belong to two categories; a whale is a mammal and an aquatic animal.

Recommendation Metrics

The RankingMetrics class can be used for evaluating recommendation models. It provides methods for quantifying the predictive effectiveness of a recommendation model.

The metrics supported by the RankingMetrics class include mean average precision, normalized discounted cumulative gain, and precision at k. You can read details about these metrics in the paper titled, "IR evaluation methods for retrieving highly relevant documents" by Kalervo Järvelin and Jaana Kekäläinen.

An Example MLlib Application

In this section, we will develop a supervised machine learning application using the MLlib API.

Dataset

We will use the Iris dataset available at https://archive.ics.uci.edu/ml/datasets/Iris. It is a widely used dataset in machine learning for testing classification algorithms.

The Iris dataset has 150 rows or observations. Each row contains the sepal length, sepal width, petal length, petal width, and species of an Iris plant. The dataset has 50 examples each for three species of Iris: Setosa, Virginica, and Versicolor.

Although, the Iris dataset is small, you can use the same code to train models with large datasets. MLlib can easily scale. It can be used to train models with datasets containing billions of observations.

Goal

The species of an Iris plant can be determined from the length and width of its sepal and petal. Different Iris species have petals and sepals of different lengths and widths. Our goal is to train a multi-nominal classifier that can predict the species of an Iris plant given its sepal length, sepal width, petal length, and petal width.

We will use sepal length, sepal width, petal length, and petal width of an Iris plant as features. The species to which a plant belongs is the label or the class of a plant.

Code

Let's launch the Spark shell from a terminal. The Spark shell makes it easy to train and use a classifier interactively.

```
path/to/spark/bin/spark-shell --master local[*]
```

Let's create an RDD from the dataset.

```
val lines = sc.textFile("/path/to/iris.data")
```

A machine learning algorithm makes several passes over a dataset, so let's cache the RDD in memory.

```
lines.persist()
```

To remove empty lines from the dataset, let's filter it.

```
val nonEmpty = lines.filter(_.nonEmpty)
```

Next, we extract the features and labels. The data is in CSV format, so let's split each line.

```
val parsed = nonEmpty map {_.split(",")}
```

The MLlib algorithms operate on RDD of LabeledPoint, so we need to transform parsed to an RDD of LabeledPoint. You may recollect that both the features and label fields in a LabeledPoint are of type Double. However, the input dataset has both the features and label in string format. Fortunately, the features are numerical values stored as strings, so converting the features to Double values is straightforward. However, the labels are stored as alphabetical strings, which need to be converted to numerical labels. To convert the name of a species to a Double typed value, we will map a species name to a number using a Map data structure. We find unique values in the species column in the dataset and assign a 0-based index to each species.

```
val distinctSpecies = parsed.map{a => a(4)}.distinct.collect
val textToNumeric = distinctSpecies.zipWithIndex.toMap
```

Now we are ready to create an RDD of LabeledPoint from parsed.

```
import org.apache.spark.mllib.regression.LabeledPoint
import org.apache.spark.mllib.linalg.{Vector, Vectors}

val labeledPoints = parsed.map{a =>
            LabeledPoint(textToNumeric(a(4)),
                Vectors.dense(a(0).toDouble, a(1).toDouble, a(2).toDouble, a(3).toDouble))}
```

Next, let's split the dataset into training and test data. We will use 80% of the data for training a model and 20% for testing it.

```
val dataSplits = labeledPoints.randomSplit(Array(0.8, 0.2))
val trainingData = dataSplits(0)
val testData = dataSplits(1)
```

Now we are ready to train a model. You can use any classification algorithm at this step. We will use the NaiveBayes algorithm.

```
import org.apache.spark.mllib.classification.NaiveBayes
val model = NaiveBayes.train(trainingData)
```

The model trained here can be used to classify an Iris plant. Given the features of an Iris plant, it can predict or tell its species.

As you can see, training a model is easy; you have to just call the train method of the algorithm that you want to use. The hard part was getting data into a format that can be used with a machine learning algorithm. This is the case for this example and in general for most machine learning applications. Generally, data scientists spend a significant amount of time and effort on data munging or wrangling activities.

Next, let's evaluate our model on the test dataset. The first step in evaluating a model is to have it predict a label for each observation in the test dataset.

```
val predictionsAndLabels = testData.map{d => (model.predict(d.features), d.label)}
```

The preceding code creates an RDD of actual and predicted labels. With this information, we can calculate various model evaluation metrics. For example, we can calculate the accuracy of the model by dividing the number of correct predictions by the number of observations in the test dataset. Alternatively, we can use the `MulticlassMetrics` class to find the precision, recall, and F-measure of our model.

```
import org.apache.spark.mllib.evaluation.MulticlassMetrics
val metrics = new MulticlassMetrics(predictionsAndLabels)

val recall = metrics.recall
```

```
recall: Double = 0.9117647058823529
```

```
val precision = metrics.precision
```

```
precision: Double = 0.9117647058823529
```

```
val fMeasure = metrics.fMeasure
```

```
fMeasure: Double = 0.9117647058823529
```

Spark ML

Spark ML is another machine learning library that runs on top of Spark. It is a relatively newer library than MLlib. It became available starting with Apache Spark version 1.2. It is also referred to as the Spark Machine Learning Pipelines API.

Spark ML provides a higher-level abstraction than MLlib for creating machine learning workflows or pipelines. It enables users to quickly assemble and tune machine learning pipelines. It makes it easy to create a pipeline for training a model, tuning a model using cross-validation, and evaluating a model with different metrics.

Many classes and singleton objects provided by the MLlib library are also provided by the Spark ML library. In fact, classes and objects related to machine learning algorithms and models have the same names in both the libraries. The classes and singleton objects provided by the Spark ML library are available under the org.apache.spark.ml package.

Generally, a machine learning task consists of the following steps:

1. Read data.

2. Preprocess or prepare data for processing.

3. Extract features.

4. Split data for training, validation, and testing.

5. Train a model with a training dataset.

6. Tune a model using cross-validation techniques.

7. Evaluate a model over a test dataset.

8. Deploy a model.

Each step represents a stage in a machine learning pipeline. Spark ML provides abstractions for these stages. Compared to MLlib, it is easier to assemble the preceding steps into a machine learning pipeline using Spark ML.

The key abstractions introduced by Spark ML include Transformer, Estimator, Pipeline, Parameter Grid, CrossValidator, and Evaluator. We will briefly describe these abstractions in this section. The next section shows concrete examples. It also shows how they are used in a machine learning application.

ML Dataset

Spark ML uses DataFrame as the primary data abstraction. Unlike in MLlib, the machine learning algorithms and models provided by Spark ML operate on DataFrames.

As discussed in Chapter 7, the DataFrame API provides a higher-level abstraction than the RDD API for representing structured data. It supports a flexible schema that allows named columns of different data types. For example, a DataFrame can have different columns storing raw data, feature vectors, actual label, and predicted label. In addition, the DataFrame API supports a wide variety of data sources.

Compared to the RDD API, the DataFrame API also makes data preprocessing and feature extraction or feature engineering easier. Data cleansing and feature engineering are generally required before a model can be fitted on a dataset. These activities constitute the majority of the work involved in a machine learning task. The DataFrame API makes it easy to generate a new column from an existing one and add it to the source DataFrame.

Transformer

A Transformer generates a new DataFrame from an existing DataFrame. It implements a method named transform, which takes a DataFrame as input and returns a new DataFrame by appending one or more new columns to the input DataFrame. A DataFrame is an immutable data structure, so a transformer does not modify the input DataFrame. Instead, it returns a new DataFrame, which includes both the columns in the input DataFrame and the new columns.

Spark ML provides two types of transformers: feature transformer and machine learning model.

Feature Transformer

A feature transformer creates one or more new columns by applying a transformation to a column in the input dataset and returns a new DataFrame with the new columns appended. For example, if the input dataset has a column containing sentences, a feature transformer can be used to split the sentences into words and create a new column that stores the words in an array.

Model

A model represents a machine learning model. It takes a DataFrame as input and outputs a new DataFrame with predicted labels for each input feature Vector. The input dataset must have a column containing feature Vectors. A model reads the column containing feature Vectors, predicts a label for each feature Vector, and returns a new DataFrame with predicted labels appended as a new column.

Estimator

An Estimator trains or fits a machine learning model on a training dataset. It represents a machine learning algorithm. It implements a method named `fit`, which takes a DataFrame as argument and returns a machine learning model.

An example of an Estimator is the `LinearRegression` class. Its `fit` method returns an instance of the `LinearRegressionModel` class.

Pipeline

A Pipeline connects multiple transformers and estimators in a specified sequence to form a machine learning workflow. Conceptually, it chains together the data preprocessing, feature extraction, and model training steps in a machine learning workflow.

A Pipeline consists of a sequence of stages, where each stage is either a Transformer or an Estimator. It runs these stages in the order they are specified.

A Pipeline itself is also an Estimator. It implements a `fit` method, which takes a DataFrame as argument and passes it through the pipeline stages. The input DataFrame is transformed by each stage. The `fit` method returns a PipelineModel, which is a Transformer.

A Pipeline's `fit` method calls the `transform` method of each Transformer and `fit` method of each Estimator in the same order as they are specified when a Pipeline is created. Each Transformer takes a DataFrame as input and returns a new DataFrame, which becomes the input for the next stage in the Pipeline. If a stage is an Estimator, its `fit` method is called to train a model. The returned model, which is a Transformer, is used to transform the output from previous stage to produce input for the next stage.

PipelineModel

A PipelineModel represents a fitted pipeline. It is generated by the `fit` method of a Pipeline. It has the same stages as the Pipeline that generated it, except for the Estimators, which are replaced by models trained by those estimators. In other words, all the Estimators are replaced by Transformers.

Unlike a Pipeline, which is an Estimator, a PipelineModel is a Transformer. It can be applied to a dataset to generate predictions for each observation. In fact, a PipelineModel is a sequence of Transformers. When the `transform` method of a PipelineModel is called with a DataFrame, it calls the `transform` method of each Transformer in sequence. Each Transformer's `transform` method outputs a new DataFrame, which becomes the input for the next Transformer in the sequence.

Evaluator

An Evaluator evaluates the predictive performance or effectiveness of a model. It provides a method named `evaluate`, which takes a DataFrame as input and returns a scalar metric. The input DataFrame passed as argument to the `evaluate` method must have columns named `label` and `prediction`.

Grid Search

The performance or quality of a machine learning model depends on the hyperparameters provided to a machine learning algorithm during model training. For example, the effectiveness of a model trained with the logistic regression algorithm depends on the step size and number of gradient descent iterations.

Unfortunately, it is difficult to pick the right combination of hyperparameters for training the best model. One of the techniques for finding the best hyperparameters is to do a grid search over a hyperparameter space. In a grid search, models are trained with each combination of hyperparameters from a specified subset of the hyperparameter space.

For example, consider a training algorithm that requires two real-valued hyperparameters: p1 and p2. Rather than guessing the best values for p1 and p2, we can do a grid search over p1 values 0.01, 0.1, and 1, and p2 values 20, 40, and 60. This results in nine different combinations of p1 and p2. A model is trained and evaluated with each combination of p1 and p2. The combination that trains a model with the best evaluation metric is selected.

Grid search is expensive, but it is a better approach for hyperparameter tuning than guessing the optimal value for each hyperparameter. Generally, it is a required step to find a model that performs well.

CrossValidator

A CrossValidator finds the best combination of hyperparameter values for training the optimal model for a machine learning task. It requires an Estimator, an Evaluator and a grid of hyperparameters.

A CrossValidator uses k-fold cross-validation and grid search for hyperparameter and model tuning. It splits a training dataset into k-folds, where k is a number specified by a user. For example, if k is 10, a CrossValidator will generate 10 pairs of training and test dataset from the input dataset. In each pair, 90% of the data is reserved for training and remaining 10% is held-out for testing.

Next, it generates all the combinations of the hyperparameters from user-specified sets of different hyperparameters. For each combination, it trains a model with a training dataset using an Estimator and evaluates the generated model over a test dataset using an Evaluator. It repeats this step for all the k-pairs of training and test datasets, and calculates the average of the specified evaluation metric for each pair.

The hyperparameters that produce the model with the best averaged evaluation metric are selected as the best hyperparameters. Finally, a CrossValidator trains a model over the entire dataset using the best hyperparameters.

Note that using a CrossValidator can be very expensive since it tries every combination of the hyperparameters from the specified parameter grid. However, it is a well-established method for choosing optimal hyperparameter values. It is statistically a better method than heuristic hand-tuning.

An Example Spark ML Application

In this section, we will develop a supervised machine learning application using the Spark ML API.

Dataset

We will use the sentiment labeled sentences dataset available at `https://archive.ics.uci.edu/ml/datasets/Sentiment+Labelled+Sentences`. It was created for the paper, "From Group to Individual Labels Using Deep Features" published by Dimitrios Kotzias, Misha Denil, Nando de Freitas, and Padhraic Smyth at KDD 2015. Download this dataset to your computer if you would like to try the code shown in this section.

This dataset contains a sample of reviews from three websites, imdb.com, amazon.com, and yelp.com. It includes randomly selected 500 positive and 500 negative reviews from each website. A review with negative sentiment is labeled 0 and positive review is labeled 1. A review is separated from its label by the tab character.

For simplicity sake, we will use only the reviews from imdb.com. These reviews are in the `imdb_labelled.txt` file.

Goal

Our goal is to train a predictive model that predicts whether a sentence has a positive or negative sentiment. To be more specific, we will train and evaluate a binary classifier using the dataset in the `imdb_labelled.txt` file.

Code

We will use the Spark shell to interactively develop a machine learning workflow. Let's launch the Spark shell from a terminal.

```
/path/to/spark/bin/spark-shell --master local[*]
```

We begin with creating an RDD from the downloaded dataset. Make sure to use the right path on your computer for the input file.

```
val lines = sc.textFile("/path/to/imdb_labelled.txt")
```

As mentioned earlier, a machine learning algorithm makes several passes over a dataset, so we can make it run faster by caching the input dataset in memory.

```
lines.persist()
```

To materialize the data in memory, let's call an action method.

```
lines.count()
```

```
res1: Long = 1000
```

Since an action triggers computation, which requires reading of data from disk, the preceding step also helps verify that the correct file path was given to the SparkContext's textFile method.

The data in its raw form cannot be directly used with a machine learning algorithm, so we need to do feature extraction or feature engineering. As the first step in the featuring engineering process, let's split the text and label for each review. As mentioned earlier, a review text and label are separated by the tab character.

```
val columns = lines.map{_.split("\\t")}
```

```
columns: org.apache.spark.rdd.RDD[Array[String]] = MapPartitionsRDD[2] at map at
<console>:23
```

The preceding code generates an RDD of Array of String. The first element in each Array is a sentence and the second element is its label. We can check an entry, as shown here.

```
columns.first
```

```
res2: Array[String] = Array("A very, very, very slow-moving, aimless movie about a
distressed, drifting young man.  ", 0)
```

Unlike MLlib, which operates on an RDD, Spark ML operates on a DataFrame. So let's create a DataFrame.

```
import sqlContext.implicits._
case class Review(text: String, label: Double)
val reviews = columns.map{a => Review(a(0),a(1).toDouble)}.toDF()
```

```
reviews: org.apache.spark.sql.DataFrame = [text: string, label: double]
```

The import statement is required to implicitly convert an RDD to a DataFrame. We created the Review case class to define the schema for our dataset. First, we create an RDD of Reviews from the RDD of Array of strings. Next, we call the toDF method, which creates a DataFrame from an RDD.

Next, as a sanity check, let's verify the schema and find the count of both positive and negative reviews.

```
reviews.printSchema
```

```
root
 |-- text: string (nullable = true)
 |-- label: double (nullable = false)
```

```
reviews.groupBy("label").count.show
```

```
+-----+-----+
|label|count|
+-----+-----+
|  1.0|  500|
|  0.0|  500|
+-----+-----+
```

To evaluate the quality or performance of a trained model, we will need a test dataset. Therefore, let's reserve a portion of the dataset for model testing.

```
val Array(trainingData, testData) = reviews.randomSplit(Array(0.8, 0.2))
trainingData.count
testData.count
```

The data is yet not in a format that can be used with a machine learning algorithm. We need to create a feature Vector for each sentence in the dataset. Spark ML provides Transformers to help with this task.

```
import org.apache.spark.ml.feature.Tokenizer
val tokenizer = new Tokenizer()
                    .setInputCol("text")
                    .setOutputCol("words")
```

Note that if you want to copy and paste this code snippet into the Spark shell, you will have to use the :paste REPL command.

Tokenizer is a Transformer that converts an input string to lowercase and splits it into words using whitespaces as separator. The tokenizer created here takes as input a DataFrame, which should have a column labeled "text", and outputs a new DataFrame, which will have all the columns in the input DataFrame and a new column labeled "words". For each sentence in the "text" column, the "words" column stores the words in an array.

We can check the schema of the DataFrame that will be output by the transform method of tokenizer, as shown next.

```
val tokenizedData = tokenizer.transform(trainingData)
```

```
tokenizedData: org.apache.spark.sql.DataFrame = [text: string, label: double,
words: array<string>]
```

Next, let's create a feature Vector to represent a sentence.

```
import org.apache.spark.ml.feature.HashingTF
val hashingTF = new HashingTF()
                    .setNumFeatures(1000)
                    .setInputCol(tokenizer.getOutputCol)
                    .setOutputCol("features")
```

HashingTF is a Transformer that converts a sequence of words into a fixed-length feature Vector. It maps a sequence of terms to their term frequencies using a hashing function.

The preceding code first creates an instance of the HashingTF class. Next, it sets the number of features, the name of the input DataFrame column that should be used for generating feature Vectors, and the name for the new column that will contain the feature Vectors generated by HashingTF.

We can check the schema of the DataFrame that will be output by the transform method of hashingTF, as shown here.

```
val hashedData = hashingTF.transform(tokenizedData)
```

```
hashedData: org.apache.spark.sql.DataFrame = [text: string, label: double, words:
array<string>, features: vector]
```

Now we have the Transformers required to transform our raw data into a format that can be used with a machine learning algorithm. The DataFrame that will be generated by the transform method of hashingTF will have a column named "label", which stores each label as a Double, and a column named "features", which stores the features for each observation as a Vector.

Next, we need an Estimator to fit a model on the training dataset. For this example, we will use the LogisticRegression class provided by the Spark ML library. It is available under the org.apache.spark.ml.classification package.

```
import org.apache.spark.ml.classification.LogisticRegression
val lr = new LogisticRegression()
                .setMaxIter(10)
                .setRegParam(0.01)
```

This code creates an instance of the LogisticRegression class and sets the maximum number of iterations and regularization parameter. Here we are trying to guess the best values for the maximum number of iterations and regularization parameter, which are the hyperparameters for the Logistic regression algorithm. We will show later how to find the optimal values for these parameters.

Now we have all the parts that we need to assemble a machine learning pipeline.

```
import org.apache.spark.ml.Pipeline
val pipeline = new Pipeline()
                    .setStages(Array(tokenizer, hashingTF, lr))
```

This code creates an instance of the Pipeline class with three stages. The first two stages are Transformers and third stage is an Estimator. The pipeline object will first use the specified Transformers to transform a DataFrame containing raw data into a DataFrame with the feature Vectors. Finally, it will use the specified Estimator to train or fit a model on the training dataset.

Now we are ready to train a model.

```
val pipeLineModel = pipeline.fit(trainingData)
```

Let's evaluate how the generated model performs on both the training and test dataset. To do this, we first get the predictions for each observation in the training and test dataset.

```
val testPredictions = pipeLineModel.transform(testData)
```

```
testPredictions: org.apache.spark.sql.DataFrame = [text: string, label: double,
words: array<string>, features: vector, rawPrediction: vector, probability: vector,
prediction: double]
```

```
val trainingPredictions = pipeLineModel.transform(trainingData)
```

```
trainingPredictions: org.apache.spark.sql.DataFrame = [text: string, label: double,
words: array<string>, features: vector, rawPrediction: vector, probability: vector,
prediction: double]
```

Note that the transform method of the pipeLineModel object created a DataFrame with three additional columns: rawPrediction, probability, and prediction.

We will use a binary classifier evaluator to evaluate our model. It expects two input columns: rawPrediction and label.

```
import org.apache.spark.ml.evaluation.BinaryClassificationEvaluator
val evaluator = new BinaryClassificationEvaluator()
```

Let's now evaluate our model using Area under ROC as a metric.

```
import org.apache.spark.ml.param.ParamMap
val evaluatorParamMap = ParamMap(evaluator.metricName -> "areaUnderROC")
```

```
val aucTraining = evaluator.evaluate(trainingPredictions, evaluatorParamMap)
```

```
aucTraining: Double = 0.9999758519725919
```

```
val aucTest = evaluator.evaluate(testPredictions, evaluatorParamMap)
```

```
aucTest: Double = 0.6984384037015618
```

Our model's AUC score is close to 1.0 for the training dataset and 0.69 for the test dataset. As mentioned earlier, a score closer to 1.0 indicates a perfect model and a score closer to 0.50 indicates a worthless model. Our model performs very well on the training dataset, but not so well on the test dataset. A model will always perform well on the dataset that it was trained with. The true performance of a model is indicated by how well it does on an unseen test dataset. That is the reason we reserved a portion of the dataset for testing.

One way to improve a model's performance is to tune its hyperparameters. Spark ML provides a CrossValidator class that can help with this task. It requires a parameter grid over which it conducts a grid search to find the best hyperparameters using k-fold cross validation.

Let's build a parameter grid that we will use with an instance of the CrossValidator class.

```
import org.apache.spark.ml.tuning.ParamGridBuilder
val paramGrid = new ParamGridBuilder()
  .addGrid(hashingTF.numFeatures, Array(10000, 100000))
  .addGrid(lr.regParam, Array(0.01, 0.1, 1.0))
  .addGrid(lr.maxIter, Array(20, 30))
  .build()
```

```
paramGrid: Array[org.apache.spark.ml.param.ParamMap] =
Array({
        logreg_0427de6fa5fc-maxIter: 20,
        hashingTF_24e660c4963c-numFeatures: 10000,
        logreg_0427de6fa5fc-regParam: 0.01
}, {
        logreg_0427de6fa5fc-maxIter: 20,
        hashingTF_24e660c4963c-numFeatures: 100000,
        logreg_0427de6fa5fc-regParam: 0.01
}, {
        logreg_0427de6fa5fc-maxIter: 20,
        hashingTF_24e660c4963c-numFeatures: 10000,
        logreg_0427de6fa5fc-regParam: 0.1
},...
```

This code created a parameter grid consisting of two values for the number of features, three values for the regularization parameters, and two values for the maximum number of iterations. It can be used to do a grid search over 12 different combinations of the hyperparameter values. You can specify more options, but training a model will take longer since grid search is a brute-force method that tries all the different combinations in a parameter grid. As mentioned earlier, using a CrossValidator to do a grid search can be expensive in terms of CPU time.

We now have all the parts required to tune the hyperparameters for the Transformers and Estimators in our machine learning pipeline.

```
import org.apache.spark.ml.tuning.CrossValidator
val crossValidator = new CrossValidator()
                        .setEstimator(pipeline)
                        .setEstimatorParamMaps(paramGrid)
```

```
            .setNumFolds(10)
            .setEvaluator(evaluator)

val crossValidatorModel = crossValidator.fit(trainingData)
```

The `fit` method in the `CrossValidator` class returns an instance of the `CrossValidatorModel` class. Similar to other model classes, it can be used as a Transformer that predicts a label for a given feature Vector.

Let's evaluate the performance of this model on the test dataset.

```
val newPredictions = crossValidatorModel.transform(testData)
```

```
newPredictions: org.apache.spark.sql.DataFrame = [text: string, label: double, words:
array<string>, features: vector, rawPrediction: vector, probability: vector, prediction:
double]
```

```
val newAucTest = evaluator.evaluate(newPredictions, evaluatorParamMap)
```

```
newAucTest: Double = 0.8182764603817234
```

As you can see, the performance of the new model is 11% better than the previous model.

Finally, let's find the best model generated by `crossValidator`.

```
val bestModel = crossValidatorModel.bestModel
```

You can use this model to classify new reviews.

Summary

Machine learning is an effective technique for building regression, classification, clustering, anomaly detection, and recommender systems. It involves training or fitting a model on historical data and using the trained model to make predictions for new data.

Spark provides two libraries for large scale machine learning: MLlib and Spark ML. Both libraries provide higher-level APIs than the core RDD API for machine learning tasks. Under the hood, both use the Spark core API. MLlib was the first machine learning library shipped with Spark. Spark ML became available starting with Spark 1.2.

Spark ML provides a higher-level abstraction than MLlib for developing machine learning pipelines or workflow. Compared to MLlib, it makes it easier to assemble a machine learning pipeline, find the best model using k-fold cross validation and grid search, and evaluate a model.

■ ■ ■

CHAPTER 9

Graph Processing with Spark

Data is generally stored and processed as a collection of records or rows. It is represented as a two-dimensional table with data divided into rows and columns. However, collections or tables are not the only way to represent data. Sometimes, a graph provides a better representation of data than a collection.

Graphs are ubiquitous. They are everywhere around us. For example, the Internet is a large graph of interconnected computers, routers, and switches. The World Wide Web is a large graph. Web pages connected by hypertext links form a graph. Social networks on sites such as Facebook, LinkedIn, and Twitter are graphs. Transportation hubs such as airports, train terminals, and bus stops can also be represented by a graph.

For graph-oriented data, a graph provides an easy-to-understand and intuitive model for working with data. In addition, specialized graph algorithms are available for processing graph-oriented data. These algorithms provide efficient tools for different analytics tasks.

This chapter introduces the library that Spark provides for efficiently processing large-scale graph-oriented data. But first let's introduce fundamental graph related terms that are used in this chapter.

Introducing Graphs

A graph is a data structure composed of vertices and edges (see Figure 9-1). A vertex is a node in graph. An edge connects two vertices in a graph. Generally, a vertex represents an entity and an edge represents a relationship between two entities. A graph is conceptually equivalent to collections of vertices and edges.

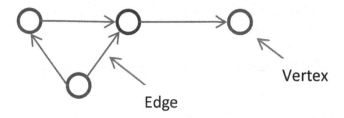

Figure 9-1. A graph

A graph can be directed or undirected.

Undirected Graphs

An undirected graph is a graph with edges that have no direction or orientation (see Figure 9-2). An edge in an undirected graph has no source or destination vertex.

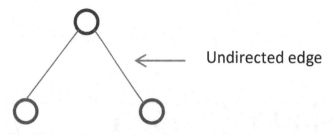

Figure 9-2. An undirected graph

Directed Graphs

A directed graph is a graph with edges that have direction or orientation (see Figure 9-3). An edge in a directed graph has a source and destination vertex.

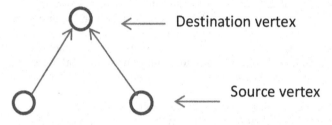

Figure 9-3. A directed graph

Directed Multigraphs

A directed multigraph is a directed graph that contains pairs of vertices connected by two or more parallel edges (see Figure 9-4). Parallel edges in a directed multigraph are edges with the same originating and terminating vertices. They are used for representing multiple relationships between a pair of vertices.

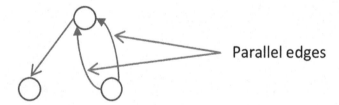

Figure 9-4. A directed multigraph

Property Graphs

A property graph is a directed multigraph that has data associated with the vertices and the edges (see Figure 9-5). Each vertex in a property graph has attributes or properties. Similarly, each edge has a label or properties.

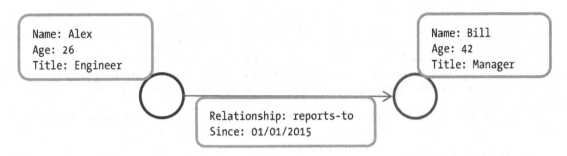

Figure 9-5. *A property graph*

A property graph provides a rich abstraction for working with graph-oriented data. It is the most popular form of modeling data with graphs.

The preceding diagram shows a property graph representing two employees in a company and their relationship.

Another example of a property graph is a graph representing a social network on Twitter. A user has attributes such as name, age, gender, and location. In addition, a user can follow other users and may have followers. In a graph representing a social network on Twitter, a vertex represents a user and an edge represents the "follows" relationship. Figure 9-6 shows a simple example.

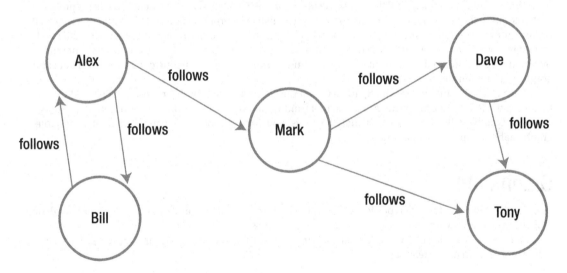

Figure 9-6. *A social network graph*

Now you know the basic graph related terms. The next section introduces GraphX.

Introducing GraphX

GraphX is a distributed graph analytics framework. It is a Spark library that extends Spark for large-scale graph processing. It provides a higher-level abstraction for graphs analytics than that provided by the Spark core API.

GraphX provides both fundamental graph operators and advanced operators implementing graph algorithms such as PageRank, strongly connected components, and triangle count. It also provides an implementation of Google's Pregel API. These operators simplify graph analytics tasks.

GraphX allows the same data to be operated on as a distributed graph or distributed collections. It provides collection operators similar to those provided by the RDD API and graph operators similar to those provided by specialized graph analytics libraries. Thus, it unifies collections and graphs as first-class composable objects.

A key benefit of using GraphX is that it provides an integrated platform for complete graph analytics workflow or pipeline. A graph analytics pipeline generally consists of the following steps:

a) Read raw data.

b) Preprocess data (e.g., cleanse data).

c) Extract vertices and edges to create a property graph.

d) Slice a subgraph.

e) Run graph algorithms.

f) Analyze the results.

g) Repeat steps e and f with another slice of the graph.

Some of these steps involve data computation and some involve graph computation. Raw data could be in XML files, database tables, CSV files, or some other format. Generally, it is not clean, so it is preprocessed first. After that, the vertices and edges are extracted to construct a property graph. This graph or a slice of it is then processed with a graph analytics library. This is a graph computation step. Preprocessing data and extracting vertices as well as edges are data computation steps. Similarly, analyzing the results generated by a graph algorithm is a data computation step. Thus, a graph analytics pipeline consists of both data and graph computation steps.

Since GraphX is integrated with Spark, a GraphX user has access to both GraphX and Spark APIs, including the RDD and DataFrame APIs. The RDD and DataFrame APIs simplify data computation steps, while the GraphX API simplifies graph analytics steps. Thus, with GraphX, you can implement a complete graph analytics pipeline with a single platform.

GraphX API

The GraphX API provides data types for representing graph-oriented data and operators for graph analytics. It allows you to work with either a graphs or a collections view of the same data.

At the time of writing of this book, GraphX is in alpha stage and undergoing rapid development. The API may change in future releases.

Data Abstractions

The key data types provided by GraphX for working with property graphs include VertexRDD, Edge, EdgeRDD, EdgeTriplet, and Graph.

VertexRDD

VertexRDD represents a distributed collection of vertices in a property graph. It provides a collection view of the vertices in a property graph. VertexRDD stores only one entry for each vertex. In addition, it indexes the entries for fast joins.

Each vertex is represented by a key-value pair, where the key is a unique id and value is the data associated with a vertex. The data type of the key is VertexId, which is essentially a 64-bit Long. The value can be of any type.

VertexRDD is a generic or parameterized class; it requires a type parameter. It is defined as VertexRDD[VD], where the type parameter *VD* specifies the data type of the attribute or property associated with each vertex in a graph. For example, *VD* can be Int, Long, Double, String or a user defined type. Thus, vertices in a graph can be represented by VertexRDD[String], VertexRDD[Int], or VertexRDD of some user defined type.

Edge

The Edge class abstracts a directed edge in a property graph. An instance of the Edge class contains source vertex id, destination vertex id and edge attributes.

Edge is also a generic class requiring a type parameter. The type parameter specifies the data type of the edge attributes. For example, an edge can have attribute of type Int, Long, Double, String or a user defined type.

EdgeRDD

EdgeRDD represents a distributed collection of the edges in a property graph. The EdgeRDD class is parameterized over the data type of the edge attributes.

EdgeTriplet

An instance of the EdgeTriplet class represents a combination of an edge and the two vertices that it connects (see Figure 9-7). It stores the attributes of an edge and the two vertices that it connects. It also contains the unique identifiers for the source and destination vertices of an edge.

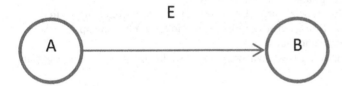

Figure 9-7. An EdgeTriplet

A collection of EdgeTriplets represents a tabular view of a property graph.

EdgeContext

The EdgeContext class combines EdgeTriplet with methods to send messages to source and destination vertices of an edge.

Graph

Graph is GraphX's abstraction for representing property graphs; an instance of the Graph class represents a property graph. Similar to RDD, it is immutable, distributed, and fault-tolerant. GraphX partitions and distributes a graph across a cluster using vertex partitioning heuristics. It recreates a partition on a different machine if a machine fails.

The Graph class unifies property graphs and collections as composable first-class objects. A property graph is conceptually equivalent to a pair of typed RDDs, where one RDD is a partitioned collection of vertices and the other one is a partitioned collection of edges. In fact, the Graph class provides access to the vertices and edges as RDDs. An RDD of vertices or edges can operated on as any other RDD. All the RDD methods are available to transform or analyze the data associated with the vertices and edges. The Graph class also provides methods to transform and analyze vertices and edges through a graph view. Thus, an instance of the Graph class can be processed as a pair of collections or a graph.

The Graph class provides methods for modifying not only the vertex and edge attributes but also the structure of a property graph. Since Graph is an immutable data type, any operator that modifies attributes or the structure returns a new property graph.

Most importantly, the Graph class provides a rich set of operators for processing and analyzing a property graph. It provides both fundamental graph operators and advanced operators that can be used to implement custom graph algorithms. It also has built-in methods that implement graph algorithms such as PageRank, strongly connected components, and triangle counting.

Creating a Graph

An instance of the Graph class can be created in several ways. The GraphX library provides an object named Graph, which provides factory methods for constructing graphs from RDDs. For example, it provides a factory method that creates an instance of the Graph class from a pair of RDDs representing vertices and edges. It provides another method that creates a graph from an RDD of edges, assigning user-provided default values to the vertices referred by the edges.

The GraphX library also provides an object named GraphLoader, which contains a method named edgeListFile. The edgeListFile method creates an instance of the Graph class from a file containing a list of directed edges. Each line must contain two integers. By default, the first integer represents the source vertex id and the second one represents the destination vertex id. The default orientation can be reversed by passing a Boolean argument to the edgeListFile method. The edgeListFile method automatically creates the vertices referred by the edges. It sets the edge and vertex attributes to 1.

Let's create a property graph representing a social network similar to Twitter's network of users. In this property graph, a vertex represents a user and a directed edge represents a "follows" relationship. Let's keep the graph simple so that it is easy to understand and play with.

A visual representation of the property graph that you will create is shown in Figure 9-8. The numbers in the vertices represent the vertex ids.

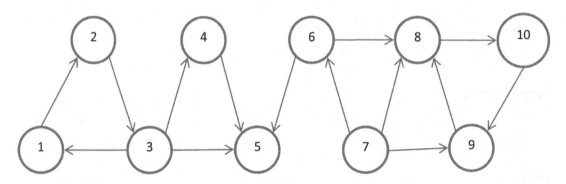

Figure 9-8. *A social network*

You can interactively explore the GraphX API using the Spark-shell. To get started, launch the Spark-shell.

```
$/path/to/spark/bin/spark-shell
```

Once you are inside the Spark-shell, import the GraphX library.

```
import org.apache.spark.graphx._
```

Next, create a case class to represent a user. Each user will have two attributes or properties: name and age.

```
case class User(name: String, age: Int)
```

Now let's create an RDD of id and user pairs.

```
val users = List((1L, User("Alex", 26)), (2L, User("Bill", 42)), (3L, User("Carol", 18)),
                 (4L, User("Dave", 16)), (5L, User("Eve", 45)), (6L, User("Farell", 30)),
                 (7L, User("Garry", 32)), (8L, User("Harry", 36)), (9L, User("Ivan", 28)),
                 (10L, User("Jill", 48))
                )
```

```
val usersRDD = sc.parallelize(users)
```

The preceding code creates an RDD of key value pairs, where the key is a unique identifier and value is a user attributes. The key represents a vertex id.

Next, let's create an RDD of connections between users. A connection can be represented by an instance of the Edge case class. An edge can have any number of attributes; however, to keep things simple, assign a single attribute of type Int to each edge.

```
val follows = List(Edge(1L, 2L, 1), Edge(2L, 3L, 1), Edge(3L, 1L, 1), Edge(3L, 4L, 1),
                   Edge(3L, 5L, 1), Edge(4L, 5L, 1), Edge(6L, 5L, 1), Edge(7L, 6L, 1),
                   Edge(6L, 8L, 1), Edge(7L, 8L, 1), Edge(7L, 9L, 1), Edge(9L, 8L, 1),
                   Edge(8L, 10L, 1), Edge(10L, 9L, 1), Edge(1L, 11L, 1)
                  )
```

```
val followsRDD = sc.parallelize(follows)
```

In the preceding code snippet, the first argument to Edge is the source vertex id. The second argument is the destination vertex id and the last argument is the edge attribute.

Note that there is an edge connecting vertex with id 1 to vertex with id 11. However, the vertex with id 11 does not have any property. GraphX allows you to handle such cases by creating a default set of properties. It will assign the default properties to the vertices that have not been explicitly assigned any properties.

```
val defaultUser = User("NA", 0)
```

Now you have all the components required to a construct a property graph.

```
val socialGraph = Graph(usersRDD, followsRDD, defaultUser)
```

```
socialGraph: org.apache.spark.graphx.Graph[User,Int] = org.apache.spark.graphx.impl.
GraphImpl@315576db
```

The Graph object in the preceding code snippet creates an instance of the Graph class from RDDs of vertices and edges.

Graph Properties

This section briefly describes how to find useful information about a property graph.

You can find the number of edges or vertices in a property graph, as shown next.

```
val numEdges = socialGraph.numEdges
```

```
numEdges: Long = 15
```

```
val numVertices = socialGraph.numVertices
```

```
numVertices: Long = 11
```

It is also easy to find the degrees of the vertices in a property graph. The degree of a vertex is the number of edges originating from or terminating at a vertex.

```
val inDegrees = socialGraph.inDegrees
```

```
inDegrees: org.apache.spark.graphx.VertexRDD[Int] = VertexRDDImpl[19] at RDD at VertexRDD.
scala:57
```

```
inDegrees.collect
```

```
res0: Array[(org.apache.spark.graphx.VertexId, Int)] = Array((4,1), (8,3), (1,1), (9,2),
(5,3), (6,1), (10,1), (2,1), (11,1), (3,1))
```

The preceding output shows that the vertex with id 4 has one incoming edge; the vertex with id 8 has three incoming edges, and so on.

```
val outDegrees = socialGraph.outDegrees
```

```
outDegrees: org.apache.spark.graphx.VertexRDD[Int] = VertexRDDImpl[23] at RDD at
VertexRDD.scala:57
```

```
outDegrees.collect
```

```
res1: Array[(org.apache.spark.graphx.VertexId, Int)] = Array((4,1), (8,1), (1,2), (9,1),
(6,2), (10,1), (2,1), (3,3), (7,3))
```

The preceding output shows that the vertex with id 4 has one outgoing edge; the vertex with id 8 has one outgoing edge, and so on.

```
val degrees = socialGraph.degrees
```

```
degrees: org.apache.spark.graphx.VertexRDD[Int] = VertexRDDImpl[27] at RDD at VertexRDD.
scala:57
```

```
degrees.collect
```

```
res2: Array[(org.apache.spark.graphx.VertexId, Int)] = Array((4,2), (8,4), (1,3), (9,3),
(5,3), (6,3), (10,2), (2,2), (11,1), (3,4), (7,3))
```

The next code snippet shows how to obtain a collections view of the vertices, edges, and triplets in a property graph.

```
val vertices = socialGraph.vertices
```

```
vertices: org.apache.spark.graphx.VertexRDD[User] = VertexRDDImpl[11] at RDD at VertexRDD.
scala:57
```

```
val edges = socialGraph.edges
```

```
edges: org.apache.spark.graphx.EdgeRDD[Int] = EdgeRDDImpl[13] at RDD at EdgeRDD.scala:40
```

```
val triplets = socialGraph.triplets
```

```
triplets: org.apache.spark.rdd.RDD[org.apache.spark.graphx.EdgeTriplet[User,Int]] =
MapPartitionsRDD[32] at mapPartitions at GraphImpl.scala:51
```

```
triplets.take(3)
```

```
res7: Array[org.apache.spark.graphx.EdgeTriplet[User,Int]] = Array(((1,User(Alex,26)),(2,U
ser(Bill,42)),1), ((2,User(Bill,42)),(3,User(Carol,18)),1), ((3,User(Carol,18)),(1,User(Al
ex,26)),1))
```

The collections view of the triplets in a property graph allows you to process a graph as a table of source vertex, destination vertex and edge properties. An example is shown next.

```
val follows = triplets.map{ t => t.srcAttr.name + " follows " + t.dstAttr.name}
```

```
follows: org.apache.spark.rdd.RDD[String] = MapPartitionsRDD[35] at map at <console>:40
```

```
follows.take(5)
```

```
res8: Array[String] = Array(Alex follows Bill, Bill follows Carol, Carol follows Alex,
Carol follows Dave, Carol follows Eve)
```

Graph Operators

This section briefly describes the key methods in the Graph class. These methods are also referred to as operators. The Graph operators can be grouped into the following categories:

- Property transformation
- Structure transformation
- Join
- Aggregation
- Graph-parallel computation
- Graph algorithm

Property Transformation Operators

The property transformation operators allow you to change the properties of the edges or vertices in a graph. Remember that Graph is immutable; a mutating operator never modifies a graph in place. Instead, it returns a new graph with the modified properties.

mapVertices

The mapVertices method applies a user-specified transformation to each vertex in a property graph. It is a higher-order method that takes a function as an argument and returns a new graph. The function provided as input to the mapVertices method takes a pair of vertex id and vertex attributes as argument and returns new vertex attributes.

The next example shows how you can use the mapVertices method to increase the age of every user in the social graph that you created earlier.

```
val updatedAges = socialGraph.mapVertices( (vertexId, user) =>
                   User(user.name, user.age + 1 ))
```

```
updatedAges: org.apache.spark.graphx.Graph[User,Int] = org.apache.spark.graphx.impl.
GraphImpl@1b8f47c6
```

```
socialGraph.vertices.take(5)
```

```
res9: Array[(org.apache.spark.graphx.VertexId, User)] = Array((4,User(Dave,16)),
(8,User(Harry,36)), (1,User(Alex,26)), (9,User(Ivan,28)), (5,User(Eve,45)))
```

```
updatedAges.vertices.take(5)
```

```
res10: Array[(org.apache.spark.graphx.VertexId, User)] = Array((4,User(Dave,17)),
(8,User(Harry,37)), (1,User(Alex,27)), (9,User(Ivan,29)), (5,User(Eve,46) ))
```

mapEdges

The mapEdges method applies a user-specified transformation to each edge in a property graph. Similar to the mapVertices method, it is a higher-order method that takes a function as an argument and returns a new graph. The function provided as input to the mapEdges method takes the attributes of an edge as argument and returns new edge attributes.

The next example modifies the label of each edge from an integer to a string.

```
val followsGraph = socialGraph.mapEdges( (n) => "follows")
```

```
followsGraph: org.apache.spark.graphx.Graph[User,String] = org.apache.spark.graphx.impl.
GraphImpl@57be3dc2
```

```
socialGraph.edges.take(5)
```

```
res11: Array[org.apache.spark.graphx.Edge[Int]] = Array(Edge(1,2,1), Edge(2,3,1),
Edge(3,1,1), Edge(3,4,1), Edge(3,5,1))
```

```
followsGraph.edges.take(5)
```

```
res12: Array[org.apache.spark.graphx.Edge[String]] = Array(Edge(1,2,follows),
Edge(2,3,follows), Edge(3,1,follows), Edge(3,4,follows), Edge(3,5,follows))
```

mapTriplets

The mapTriplets method applies a user-specified transformation to each triplet in a property graph. It is a higher-order method that takes a function as an argument and returns a new graph.

The mapTriplets method is similar to the mapEdges method; both return a new graph by modifying the properties of each edge in the source graph. However, the input function provided to the mapTriplets method takes an EdgeTriplet as an argument, whereas the function argument to the mapEdges method takes an Edge as an argument.

Every edge in the example social graph has an integer attribute whose value is always 1. Suppose you wanted to change the label of the edges based on the source and destination vertices. The following code snippet changes an edge label to 2 if the age of a follower is greater than 30.

```
val weightedGraph = socialGraph.mapTriplets{ t =>
                                  if (t.srcAttr.age >= 30)
                                    2
                                  else
                                    1
                              }
```

```
weightedGraph: org.apache.spark.graphx.Graph[User,Int] = org.apache.spark.graphx.impl.
GraphImpl@15c484a1
```

```
socialGraph.edges.take(10)
```

```
res15: Array[org.apache.spark.graphx.Edge[Int]] = Array(Edge(1,2,1), Edge(2,3,1),
Edge(3,1,1), Edge(3,4,1), Edge(3,5,1), Edge(4,5,1), Edge(6,5,1), Edge(6,8,1), Edge(7,6,1),
Edge(7,8,1))
```

```
weightedGraph.edges.take(10)
```

```
res16: Array[org.apache.spark.graphx.Edge[Int]] = Array(Edge(1,2,1), Edge(2,3,2),
Edge(3,1,1), Edge(3,4,1), Edge(3,5,1), Edge(4,5,1), Edge(6,5,2), Edge(6,8,2), Edge(7,6,2),
Edge(7,8,2))
```

Structure Transformation Operators

The operators in this category return a graph that is structurally different from the graph on which the operator was called. Conceptually, they change the structure of a graph or return a subgraph. However, remember that Graph is immutable; these operators create a new graph.

reverse

The reverse method reverses the direction of all the edges in a property graph. It returns a new property graph.

The following code snippet reverses the follow relationship in the example social graph.

```
val reverseGraph = socialGraph.reverse
```

```
reverseGraph.triplets.map{ t => t.srcAttr.name + " follows " + t.dstAttr.name}.take(10)
```

```
res21: Array[String] = Array(Alex follows Carol, Bill follows Alex, Carol follows Bill,
Dave follows Carol, Eve follows Carol, Eve follows Dave, Eve follows Farell, Farell
follows Garry, Harry follows Farell, Harry follows Garry)
```

```
socialGraph.triplets.map{ t => t.srcAttr.name + " follows " + t.dstAttr.name}.take(10)
```

```
res22: Array[String] = Array(Alex follows Bill, Bill follows Carol, Carol follows Alex,
Carol follows Dave, Carol follows Eve, Dave follows Eve, Farell follows Eve, Farell
follows Harry, Garry follows Farell, Garry follows Harry)
```

subgraph

The subgraph method applies a user-specified filter to each vertex and edge. It returns a subgraph of the source graph. Conceptually, it is similar to the RDD filter method.

The subgraph method takes two predicates (functions returning Boolean results) as arguments. The first predicate takes an EdgeTriplet as argument. The second predicate takes a pair of vertex id and vertex property as arguments. The returned subgraph contains only the vertices and edges that satisfy the two predicates.

The following code snippet creates a subgraph that contains only edges with weight greater than 1.

```
val subgraph = weightedGraph.subgraph( edgeTriplet => edgeTriplet.attr > 1,
                                       (vertexId, vertexProperty) => true)

subgraph.edges.take(10)
```

```
res23: Array[org.apache.spark.graphx.Edge[Int]] = Array(Edge(2,3,2), Edge(6,5,2),
Edge(6,8,2), Edge(7,6,2), Edge(7,8,2), Edge(7,9,2), Edge(8,10,2), Edge(10,9,2))
```

```
weightedGraph.edges.take(10)
```

```
res24: Array[org.apache.spark.graphx.Edge[Int]] = Array(Edge(1,2,1), Edge(2,3,2),
Edge(3,1,1), Edge(3,4,1), Edge(3,5,1), Edge(4,5,1), Edge(6,5,2), Edge(6,8,2), Edge(7,6,2),
Edge(7,8,2))
```

mask

The mask method takes a graph as argument and returns a subgraph of the source graph containing all the vertices and edges in the input graph. The returned graph contains vertices and edges that exist in both the input graph and source graph; however, the vertex and edge properties in the returned graph are from the source graph.

The following example uses the mask method to obtain a subgraph of the social graph. It is a contrived example that assumes that you have a list of edges, but do not have the properties of the edges or vertices connected by those edges.

```
val femaleConnections = List(Edge(2L, 3L, 0), Edge(3L, 1L, 0), Edge(3L, 4L, 0),
                             Edge(3L, 5L, 0), Edge(4L, 5L, 0), Edge(6L, 5L, 0),
                             Edge(8L, 10L, 0), Edge(10L, 9L, 0)
                            )

val femaleConnectionsRDD =  sc.parallelize(femaleConnections)

val femaleGraphMask = Graph.fromEdges(femaleConnectionsRDD, defaultUser)
```

```
femaleGraphMask: org.apache.spark.graphx.Graph[User,Int] = org.apache.spark.graphx.impl.
GraphImpl@9df22e0
```

```
val femaleGraph = socialGraph.mask(femaleGraphMask)
```

```
femaleGraph: org.apache.spark.graphx.Graph[User,Int] = org.apache.spark.graphx.impl.
GraphImpl@38c13d96
```

```
femaleGraphMask.triplets.take(10)
```

```
res31: Array[org.apache.spark.graphx.EdgeTriplet[User,Int]] = Array(((2,User(NA,0)),
(3,User(NA,0)),0), ((3,User(NA,0)),(1,User(NA,0)),0), ((3,User(NA,0)),(4,User(NA,0)),0),
((3,User(NA,0)),(5,User(NA,0)),0), ((4,User(NA,0)),(5,User(NA,0)),0), ((6,User(NA,0)),
(5,User(NA,0)),0), ((8,User(NA,0)),(10,User(NA,0)),0), ((10,User(NA,0)),(9,User(NA,0)),0))
```

```
femaleGraph.triplets.take(10)
```

```
res32: Array[org.apache.spark.graphx.EdgeTriplet[User,Int]] = Array(((2,User(Bill,42)),(3
,User(Carol,18)),1), ((3,User(Carol,18)),(1,User(Alex,26)),1), ((3,User(Carol,18)),(4,Use
r(Dave,16)),1), ((3,User(Carol,18)),(5,User(Eve,45)),1), ((8,User(Harry,36)),(10,User(Ji
ll,48)),1), ((10,User(Jill,48)),(9,User(Ivan,28)),1))
```

groupEdges

The groupEdges method is a higher-order method that merges parallel edges in a property graph. It takes a function as argument and returns a new graph. The input function takes a pair of edge properties as arguments, merges them, and returns a new property.

The following example code reduces the size of a graph by merging parallel edges connecting the same pairs of vertices.

```
val multiEdges = List(Edge(1L, 2L, 100), Edge(1L, 2L, 200),
                      Edge(2L, 3L, 300), Edge(2L, 3L, 400),
                      Edge(3L, 1L, 200), Edge(3L, 1L, 300)
                 )
```

```
val multiEdgesRDD = sc.parallelize(multiEdges)
```

```
val defaultVertexProperty = 1
```

```
val multiEdgeGraph = Graph.fromEdges(multiEdgesRDD, defaultVertexProperty)
```

```
import org.apache.spark.graphx.PartitionStrategy._
```

```
val repartitionedGraph = multiEdgeGraph.partitionBy(CanonicalRandomVertexCut)
```

```
val singleEdgeGraph = repartitionedGraph.groupEdges((edge1, edge2) => edge1 + edge2)
```

```
singleEdgeGraph: org.apache.spark.graphx.Graph[Int,Int] = org.apache.spark.graphx.impl.
GraphImpl@234a8efe
```

```
multiEdgeGraph.edges.collect
```

```
res44: Array[org.apache.spark.graphx.Edge[Int]] = Array(Edge(1,2,100), Edge(1,2,200),
Edge(2,3,300), Edge(2,3,400), Edge(3,1,200), Edge(3,1,300))
```

```
singleEdgeGraph.edges.collect
```

```
res45: Array[org.apache.spark.graphx.Edge[Int]] = Array(Edge(3,1,500), Edge(1,2,300),
Edge(2,3,700))
```

Note that the groupEdges method requires the parallel edges to be co-located on the same partition. Therefore, you must call partitionBy before calling groupEdges; otherwise, you will get incorrect results.

The partitionBy method repartitions the edges in the source graph according to a user-specified partition strategy. The preceding example used CanonicalRandomVertexCut, which assigns edges to partitions by hashing the source and destination vertex ids. It co-locates all the edges between two vertices.

Join Operators

The join operators allow you to update existing properties or add a new property to the vertices in a graph.

joinVertices

The joinVertices method updates the vertices in the source graph with a collection of vertices provided to it as input. It takes two arguments. The first argument is an RDD of vertex id and vertex data pairs. The second argument is a user-specified function that updates a vertex in the source graph using its current properties and new input data. Vertices that do not have entries in the input RDD retain their current properties.

Suppose that you found that the ages of users with vertex id 3 and 4 in the example social graph are incorrect. The following code example uses the joinVertices method to correct the ages of those two users.

```
val correctAges = sc.parallelize(List((3L, 28), (4L, 26)))
val correctedGraph = socialGraph.joinVertices(correctAges)((id, user, correctAge) =>
                                User(user.name, correctAge))
```

```
correctedGraph: org.apache.spark.graphx.Graph[User,Int] = org.apache.spark.graphx.impl.
GraphImpl@f3cb93
```

```
correctedGraph.vertices.collect
```

```
res53: Array[(org.apache.spark.graphx.VertexId, User)] = Array((4,User(Dave,26)),
(8,User(Harry,36)), (1,User(Alex,26)), (9,User(Ivan,28)), (5,User(Eve,45)),
(6,User(Farell,30)), (10,User(Jill,48)), (2,User(Bill,42)), (11,User(NA,0)),
(3,User(Carol,28)), (7,User(Garry,32)))
```

```
socialGraph.vertices.collect
```

```
res54: Array[(org.apache.spark.graphx.VertexId, User)] = Array((4,User(Dave,16)),
(8,User(Harry,36)), (1,User(Alex,26)), (9,User(Ivan,28)), (5,User(Eve,45)),
(6,User(Farell,30)), (10,User(Jill,48)), (2,User(Bill,42)), (11,User(NA,0)),
(3,User(Carol,18)), (7,User(Garry,32)))
```

The following code snippet makes it easy to compare the properties for the vertices with id 3 and 4 in socialGraph and correctedGraph.

```
val incorrectSubGraph = socialGraph.subgraph( edgeTriplet => true,
          (vertexId, vertexProperty) => (vertexId == 3) || (vertexId == 4))
```

```
val correctedSubGraph = correctedGraph.subgraph( edgeTriplet => true,
          (vertexId, vertexProperty) => (vertexId == 3) || (vertexId == 4))
```

```
incorrectSubGraph.vertices.collect
```

```
res50: Array[(org.apache.spark.graphx.VertexId, User)] = Array((4,User(Dave,16)),
(3,User(Carol,18)))
```

```
correctedSubGraph.vertices.collect
```

```
res51: Array[(org.apache.spark.graphx.VertexId, User)] = Array((4,User(Dave,26)),
(3,User(Carol,28)))
```

outerJoinVertices

The outerJoinVertices method adds new properties to the vertices in the source graph. Similar to the joinVertices method, it takes two arguments. The first argument is an RDD of pairs of vertex id and vertex data. The second argument is a user-specified function that takes as argument a triple containing a vertex id, current vertex attributes, and optional attributes in the input RDD for the same vertex. It returns new attributes for each vertex.

The input RDD should contain at most one entry for each vertex in the source graph. If the input RDD does not have an entry for a vertex in the source graph, the user-specified function receives None as the third argument when it is called on that vertex.

The difference between joinVertices and outerJoinVertices is that the former updates current attributes, while the latter can both update current attributes and add new attributes to the vertices in the source graph.

Suppose you wanted to add a new property called city to each user in the example social graph. The following code example uses the outerJoinVertices method to create a new property graph, where each user has a name, age, and city.

```
case class UserWithCity(name: String, age: Int, city: String)
```

```
val userCities = sc.parallelize(List((1L, "Boston"), (3L, "New York"), (5L, "London"),
                                (7L, "Bombay"), (9L, "Tokyo"), (10L, "Palo Alto")))
```

```
val  socialGraphWithCity = socialGraph.outerJoinVertices(userCities)((id, user, cityOpt) =>
                         cityOpt match {
                             case Some(city) => UserWithCity(user.name, user.age, city)
                             case None => UserWithCity(user.name, user.age, "NA")
                         })
```

```
socialGraphWithCity: org.apache.spark.graphx.Graph[UserWithCity,Int] = org.apache.spark.
graphx.impl.GraphImpl@b5a2596
```

```
socialGraphWithCity.vertices.take(5)
```

```
res55: Array[(org.apache.spark.graphx.VertexId, UserWithCity)] = Array((4,UserWithCity
(Dave,16,NA)), (8,UserWithCity(Harry,36,NA)), (1,UserWithCity(Alex,26,Boston)),
(9,UserWithCity(Ivan,28,Tokyo)), (5,UserWithCity(Eve,45,London)))
```

Aggregation Operators

This category currently has only one method, which is described next.

aggregateMessages

The aggregateMessages method aggregates values for each vertex from neighboring vertices and connecting edges. It uses two user-defined functions to do the aggregations. It calls these functions for every triplet in a property graph.

The first user-defined function takes as argument an EdgeContext. The EdgeContext class provides not only access to the properties of the source and destination vertices in a triplet as well as the edge connecting them, but also functions to send messages to the source and destination vertices in a triplet. A message is essentially some data. The user-defined function sends messages to neighboring vertices using the message sending function provided by the EdgeContext class.

The second user-defined function takes two messages as arguments and returns a single message. It merges or combines two messages into a single message. The aggregateMessages method uses the second function to aggregate all the messages sent by the first function to a vertex. It returns an RDD of vertex id and aggregated message pairs.

The following code shows how you can use the aggregateMessages method to calculate the number of followers of each user in the example social graph.

```
val followers = socialGraph.aggregateMessages[Int]( edgeContext => edgeContext.sendToDst(1),
                         (x, y) => (x + y))
```

```
followers: org.apache.spark.graphx.VertexRDD[Int] = VertexRDDImpl[231] at RDD at
VertexRDD.scala:57
```

```
followers.collect
```

```
res56: Array[(org.apache.spark.graphx.VertexId, Int)] = Array((4,1), (8,3), (1,1), (9,2),
(5,3), (6,1), (10,1), (2,1), (11,1), (3,1))
```

This code is a simple example for demonstrating how you can use the aggregateMessages method. However, the aggregateMessages method is overkill for calculating the number of followers. A simpler approach is shown next.

```
val followers = socialGraph.inDegrees
```

```
followers: org.apache.spark.graphx.VertexRDD[Int] = VertexRDDImpl[19] at RDD at VertexRDD.
scala:57
```

```
followers.collect
```

```
res57: Array[(org.apache.spark.graphx.VertexId, Int)] = Array((4,1), (8,3), (1,1), (9,2),
(5,3), (6,1), (10,1), (2,1), (11,1), (3,1))
```

Let's use the aggregateMessages method to solve a slightly more difficult problem. The following code snippet shows how you can find the age of the oldest follower of each user.

```
val oldestFollower = socialGraph.aggregateMessages[Int]( edgeContext =>
                           edgeContext.sendToDst(edgeContext.srcAttr.age),
                           (x, y) => math.max(x,y))
oldestFollower.collect
```

```
res58: Array[(org.apache.spark.graphx.VertexId, Int)] = Array((4,18), (8,32), (1,18),
(9,48), (5,30), (6,32), (10,36), (2,26), (11,26), (3,42))
```

Graph-Parallel Operators

The graph-parallel operators allow you to implement custom iterative graph algorithms for processing large-scale graph-oriented data.

pregel

The pregel method implements a variant of Google's Pregel algorithm. Pregel is a distributed computing model for processing large-scale graphs with billions of vertices and trillions of edges. It is based on the Bulk Synchronous Parallel (BSP) algorithm.

In Pregel, a graph computation consists of a sequence of iterations, also known as supersteps. During a superstep, a user-defined function, also known as vertex program, is invoked on each vertex in parallel. A vertex program processes the aggregate of the messages sent to a vertex in the previous superstep, updates the properties of a vertex and its outgoing edges, and sends messages to other vertices. The messages sent during a superstep are processed in the next superstep. If no messages are sent to a vertex in a superstep, the vertex program will not be invoked on it in the next superstep. All the vertex programs are allowed to finish processing messages from the previous superstep before the framework runs the next iteration of the algorithm. These iterations or supersteps are repeated until there are no messages in transit and all the vertices are inactive. A vertex becomes inactive when it has no further work to do. It gets reactivated by a message sent to it by another vertex.

GraphX's Pregel implementation is slightly different from standard implementations. The vertex program called in each superstep takes an `EdgeTriplet` as one of the arguments. It has access to the properties of the source and destination vertices in an `EdgeTriplet` as well as the edge connecting them. In addition, GraphX allows messages to be sent only to neighboring vertices.

The `pregel` method takes two sets of arguments and returns a graph. The first set contains three arguments: the initial message that each vertex will receive in the first superstep, the maximum number of iterations, and the direction in which messages will be sent.

The second set also contains three arguments: a user-defined vertex program that will be run in parallel on each vertex during a superstep, a user-defined function that computes optional messages for the neighboring vertices, and a user-defined function that merges or combines messages sent to a vertex in the previous superstep.

The user-defined vertex program takes three arguments: the id of a vertex, its current properties, and an aggregation of all the messages sent to it in the previous superstep. It returns the new properties for a vertex. In the first superstep or iteration, all the vertices receive the initial message passed as an argument to the `pregel` method. In subsequent iterations, each vertex receives an aggregate of all the messages sent to it in the previous superstep or iteration. If a vertex did not receive a message, the vertex program is not invoked on it.

The `pregel` method invokes the user-defined send message function on all the outgoing edges after it has run the vertex program in parallel on all the vertices. The send message function computes the optional messages for the neighboring vertices. It takes an `EdgeTriplet` as argument and returns an iterator of destination vertex id and message pairs. This function is invoked only on the edges of the vertices that received messages in a superstep.

The user-defined message combiner function takes a pair of messages as arguments, merges them, and returns a single message of the same type as the incoming messages. It must be a commutative and associative function.

The `pregel` method iterates until there are no messages or it reaches the user-specified maximum iterations. It returns the resulting graph at the end of the computation.

The `pregel` operator is useful for graph-parallel computations that recursively compute vertex properties. In such computations, the properties of a vertex depend on the properties of the neighboring vertices, which in turn depend on their neighboring vertices, and so on. These algorithms are iterative in nature. The computations in each iteration depend on the results of the previous iteration. The properties of the vertices are recomputed in each iteration until a stop-criteria is met.

An example of a graph-parallel computation is the PageRank algorithm, the original algorithm used by the Google search engine. A detailed discussion of the PageRank algorithm is out of scope for this book, but it is briefly described here since the next example uses the `pregel` operator to implement PageRank.

The PageRank of a document indicates the importance of a document. A document with a higher PageRank is more important than a page with a lower PageRank. The PageRank of a document is based on the PageRank of the documents linking to it. Thus, it is an iterative algorithm. In addition to the PageRank, the number of outgoing links from a document is also a factor in calculating the contribution of a document to the PageRank of a document that it links to. A document with a large number of outgoing links contributes less to the PageRank of a linked document compared to a page with the same PageRank but fewer outgoing links.

Let's use the `pregel` method and PageRank algorithm to calculate the influence rank of a user in the example social graph. A user's influence rank is based on the influence rank of the followers and the number of people followed by each follower.

The following example uses the same social graph that was used in the previous examples. To use the `pregel` operator, you need to make some changes to the social graph, represented by `socialGraph`. Currently, every edge has the same attribute. Instead, assign a weight to each edge based on the number of outgoing edges from a vertex.

```
val outDegrees = socialGraph.outDegrees
```

```
outDegrees: org.apache.spark.graphx.VertexRDD[Int] = VertexRDDImpl[23] at RDD at
VertexRDD.scala:57
```

```
val outDegreesGraph = socialGraph.outerJoinVertices(outDegrees) {
                         (vId, vData, OptOutDegree) =>
                             OptOutDegree.getOrElse(0)
               }
```

```
outDegreesGraph: org.apache.spark.graphx.Graph[Int,Int] = org.apache.spark.graphx.impl.
GraphImpl@75cffddf
```

```
val weightedEdgesGraph = outDegreesGraph.mapTriplets{EdgeTriplet =>
                               1.0 / EdgeTriplet.srcAttr
                         }
```

```
weightedEdgesGraph: org.apache.spark.graphx.Graph[Int,Double] = org.apache.spark.graphx.
impl.GraphImpl@95cd375
```

Next, assign an initial influence rank to each user.

```
val inputGraph = weightedEdgesGraph.mapVertices((id, vData) => 1.0)
```

You will be calling the pregel method on the inputGraph. However, before you do that, let's initialize the arguments that will be passed to the pregel method.

The following code snippet shows the first set of three arguments that you will pass to the pregel method.

```
val firstMessage = 0.0
```

```
val iterations = 20
```

```
val edgeDirection = EdgeDirection.Out
```

Next, define the three functions that will be passed as arguments to the pregel method.

```
val updateVertex = (vId: Long, vData: Double, msgSum: Double) => 0.15 + 0.85 * msgSum
```

```
updateVertex: (Long, Double, Double) => Double = <function3>
```

```
val sendMsg = (triplet: EdgeTriplet[Double, Double]) =>
                 Iterator((triplet.dstId, triplet.srcAttr * triplet.attr))
```

```
sendMsg: org.apache.spark.graphx.EdgeTriplet[Double,Double] => Iterator[(org.apache.spark.
graphx.VertexId, Double)] = <function1>
```

```
val aggregateMsgs = (x: Double, y: Double ) => x + y
```

```
aggregateMsgs: (Double, Double) => Double = <function2>
```

Now let's call the pregel method.

```
val influenceGraph = inputGraph.pregel(firstMessage, iterations, edgeDirection)(updateVertex,
                                        sendMsg, aggregateMsgs)
```

```
influenceGraph: org.apache.spark.graphx.Graph[Double,Double] = org.apache.spark.graphx.
impl.GraphImpl@ad6a671
```

Let's check the result of the computation by printing the name and influence rank of each user.

```
val userNames = socialGraph.mapVertices{(vId, vData) => vData.name}.vertices
```

```
val userNamesAndRanks = influenceGraph.outerJoinVertices(userNames) {
                            (vId, rank, optUserName) =>
                                     (optUserName.get, rank)
                            }.vertices
```

```
userNamesAndRanks.collect.foreach{ case(vId, vData) =>
                            println(vData._1 +"'s influence rank: " + vData._2)
                            }
```

```
Dave's influence rank: 0.2546939612521768
Harry's influence rank: 0.9733813220346056
Alex's influence rank: 0.2546939612521768
Ivan's influence rank: 0.9670543985781628
Eve's influence rank: 0.47118379300959834
Farell's influence rank: 0.1925
Jill's influence rank: 0.9718993399637271
Bill's influence rank: 0.25824491587871073
NA's influence rank: 0.25824491587871073
Carol's influence rank: 0.36950816084343974
Garry's influence rank: 0.15
```

Graph Algorithms

The aggregateMessages and pregel operators are powerful tools for implementing custom graph algorithms. Complex graph algorithms can be implemented with a few lines of code using these operators. However, in some cases, even that is not required. GraphX comes with built-in implementations for a few common graph algorithms. These algorithms can be invoked as method calls on the Graph class.

pageRank

The pageRank method implements a dynamic version of the PageRank algorithm. It takes two arguments. The first argument is the convergence threshold or tolerance. The dynamic PageRank algorithm runs until the ranks stop changing by more than a user specified tolerance. The second argument is optional. It specifies the random reset probability used by the PageRank algorithm. The default value is 0.15. The pageRank method returns a graph with PageRank as vertex attribute and normalized weight as edge attribute.

An example is shown next. It runs the dynamic PageRank on the inputGraph that was created earlier to demonstrate the pregel method.

```
val dynamicRanksGraph = inputGraph.pageRank(0.001)

dynamicRanksGraph.vertices.collect
```

```
res63: Array[(org.apache.spark.graphx.VertexId, Double)] = Array((4,0.2544208826877713),
(8,1.4102365771190497), (1,0.2544208826877713), (9,1.3372775106440455),
(5,0.5517909837066515), (6,0.1925), (10,1.3487010905511918), (2,0.2577788005094401),
(11,0.2577788005094401), (3,0.36854429183919274), (7,0.15)))
```

```
dynamicRanksGraph.edges.collect
```

```
res64: Array[org.apache.spark.graphx.Edge[Double]] = Array(Edge(1,2,0.5), Edge(2,3,1.0),
Edge(3,1,0.3333333333333333), Edge(3,4,0.3333333333333333), Edge(3,5,0.3333333333333333),
Edge(4,5,1.0), Edge(6,5,0.5), Edge(6,8,0.5), Edge(7,6,0.3333333333333333),
Edge(7,8,0.3333333333333333), Edge(7,9,0.3333333333333333), Edge(1,11,0.5),
Edge(8,10,1.0), Edge(9,8,1.0), Edge(10,9,1.0))
```

staticPageRank

The staticPageRank method implements a static version of the PageRank algorithm. It takes two arguments. The first argument is the number of iterations and the second optional argument is the random reset probability. The staticPageRank method runs the specified number of iterations and returns a graph with PageRank as vertex attribute and normalized weight as edge attribute.

```
val staticRanksGraph = inputGraph.staticPageRank(20)

staticRanksGraph.vertices.collect
```

```
res66: Array[(org.apache.spark.graphx.VertexId, Double)] = Array((4,0.2546939612521768),
(8,1.3691322478088301), (1,0.2546939612521768), (9,1.2938371623789482),
(5,0.5529962930095983), (6,0.1925), (10,1.3063496503175922), (2,0.25824491587871073),
(11,0.25824491587871073), (3,0.36950816084343974), (7,0.15)))
```

```
staticRanksGraph.edges.collect
```

```
res67: Array[org.apache.spark.graphx.Edge[Double]] = Array(Edge(1,2,0.5), Edge(2,3,1.0),
Edge(3,1,0.3333333333333333), Edge(3,4,0.3333333333333333), Edge(3,5,0.3333333333333333),
Edge(4,5,1.0), Edge(6,5,0.5), Edge(6,8,0.5), Edge(7,6,0.3333333333333333),
Edge(7,8,0.3333333333333333), Edge(7,9,0.3333333333333333), Edge(1,11,0.5),
Edge(8,10,1.0), Edge(9,8,1.0), Edge(10,9,1.0))
```

connectedComponents

The connectedComponents method implements the connected components algorithm, which computes the connected component membership of each vertex. It returns a graph, where each vertex's property is the vertex id of the lowest-numbered vertex in a connected component.

```
val connectedComponentsGraph = inputGraph.connectedComponents()
```

```
connectedComponentsGraph.vertices.collect
```

```
res68: Array[(org.apache.spark.graphx.VertexId, org.apache.spark.graphx.VertexId)] =
Array((4,1), (8,1), (1,1), (9,1), (5,1), (6,1), (10,1), (2,1), (11,1), (3,1), (7,1))
```

stronglyConnectedComponents

A strongly connected component (SCC) of a graph is a subgraph containing vertices that are reachable from every other vertex in the same subgraph. In a SCC, any vertex can be reached from any other vertex by traversing the edges in the directions in which they point (see Figure 9-9).

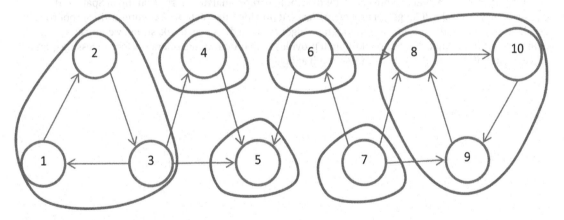

Figure 9-9. Strongly connected components

The stronglyConnectedComponents method finds the SCC for each vertex and returns a graph, where a vertex's property is the lowest vertex id in the SCC containing that vertex.

```
val sccGraph = inputGraph.stronglyConnectedComponents(20)
sccGraph.vertices.collect
```

```
res69: Array[(org.apache.spark.graphx.VertexId, org.apache.spark.graphx.VertexId)] =
Array((4,4), (8,8), (1,1), (9,8), (5,5), (6,6), (10,8), (2,1), (11,11), (3,1), (7,7))
```

triangleCount

The triangleCount method computes for each vertex the number of triangles containing it. A vertex is a part of a triangle if it has two neighbor vertices with an edge between them. The triangleCount method returns a graph, where a vertex's property is the number of triangles containing it.

```
val triangleCountGraph = inputGraph.triangleCount()
triangleCountGraph.vertices.collect
```

```
res70: Array[(org.apache.spark.graphx.VertexId, Int)] = Array((4,1), (8,3), (1,1), (9,2),
(5,1), (6,1), (10,1), (2,1), (11,0), (3,2), (7,2))
```

Summary

Data can be sometimes better represented by a graph view than a collection view. For such graph-oriented data, graph algorithms provide better tools for analytics than the algorithms that operate on the collections view of data.

GraphX is a library that extends Spark for distributed graph analytics. It runs on top of Spark and exposes a higher-level API for graph processing tasks. It provides abstractions for representing property graphs and comes with built-in support for graph algorithms such as PageRank, strongly connect components, and triangle count. In addition, it provides both fundamental operators for processing graphs and a variant of the Pregel API for complex graph analytics.

CHAPTER 10

■ ■ ■

Cluster Managers

A cluster manager manages a cluster of computers. To be more specific, it manages resources such as CPU, memory, storage, ports, and other resources available on a cluster of nodes. It pools together the resources available on each cluster node and enables different applications to share these resources. Thus, it turns a cluster of commodity computers into a virtual super-computer that can be shared by multiple applications.

Distributed computing frameworks use either an embedded cluster manager or an external cluster manager. For example, prior to version 2.0, Hadoop MapReduce used an embedded cluster manager. In Hadoop 2.0, the cluster manager was separated from the compute engine. It became an independent component that could be paired with any compute engine, including MapReduce, Spark, and Tez.

Spark comes prepackaged with a cluster manager, but it can also be used with a few other open source cluster managers. This chapter describes the different cluster managers supported by Spark.

Note that the Spark application programming interface is independent of the cluster managers; a Spark application can be deployed with any of the cluster managers supported by Spark without requiring any code change. Thus, a Spark application is *cluster manager agnostic*.

Spark supports three cluster managers: Apache Mesos, Hadoop YARN, and the Standalone cluster manager. It provides a script that can be used to deploy a Spark application with any of the supported cluster managers.

Standalone Cluster Manager

The Standalone cluster manager comes prepackaged with Spark. It offers the easiest way to set up a Spark cluster.

Architecture

The Standalone cluster manager consists of two key components: master and worker (see Figure 10-1). The worker process manages the compute resources on a cluster node. The master process pools together the compute resources from all the workers and allocates them to applications. The master process can be run on a separate server or it can run on one of the worker nodes along with a worker process.

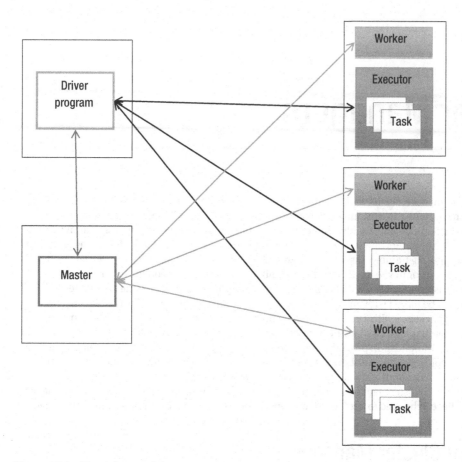

Figure 10-1. *Standalone architecture*

A Spark application deployed with the Standalone cluster manager involves the following key entities: driver program, executor, and task. A driver program is an application that uses the Spark library. It provides the data processing logic. An executor is a JVM process running on a worker node. It executes the data processing jobs submitted by a driver program. An executor can run one or more tasks simultaneously. It also provides memory for caching data.

A driver program connects to a cluster using a SparkContext object, which is the entry point into the Spark library. The SparkContext object uses the master to acquire compute resources on the worker nodes. Next, it launches executors on the worker nodes and sends application code to the executors. Finally, it splits a data processing job into tasks and sends them to the executors for execution.

Setting Up a Standalone Cluster

Setting up a standalone cluster is simple; just download and extract the Spark binaries on each cluster node. Next, start the master and the workers. The master and workers can be started manually one at a time or all of them can be started simultaneously with a script provided by Spark.

Manually Starting a Cluster

First, start the master process on one of the servers and then launch a worker process on each cluster node. The master process can be started, as shown next.

```
$ /path/to/spark/sbin/start-master.sh
```

By default, the master process listens for incoming connections on a non-local IP address on port 7077. Both the IP address and port are configurable.

Next, start a worker on each node in your Spark cluster. It can be launched, as shown next.

```
$ /path/to/spark/sbin/start-slave.sh <master-URL>
```

The master-URL is in the form spark://HOST:PORT, where HOST is the hostname or IP address of the machine running the master process and PORT is the TCP port on which the master is listening.

The scripts for launching the master and workers accept command line arguments. The commonly used arguments are listed in Table 10-1.

Table 10-1. *Command-Line Options for the Standalone Master and Worker*

Option	Description
--h HOST	Hostname or IP address on which a master or worker listens on.
--p PORT	TCP port on which a master or worker listens on. By default, a master listens on port 7077 and a worker listens on a random port.
--webui-port PORT	TCP port for monitoring a master or worker from a browser. The default port is 8080 for a master and 8081 for the workers.
--cores CORES	The number of CPU cores that a worker should allocate to Spark applications. By default, a worker allocates all cores.
--memory MEM	The amount of memory that a worker should allocate to Spark applications. By default, a worker allocates total system memory minus 1 GB.

Once the master and worker processes have been started, a Spark cluster is ready to run Spark applications. You can check the status of a cluster by connecting to the master on its webUI port (default 8080) from a browser. Chapter 11 describes the web-based monitoring interface provided by Spark.

Manually Stopping a Cluster

A Spark worker can be stopped using a script provided by Spark, as shown next.

```
$ /path/to/spark/sbin/stop-slave.sh
```

You have to do this on every worker machine.
A Spark master can also be stopped using a script provided by Spark.

```
$ /path/to/spark/sbin/stop-master.sh
```

Starting a Cluster with a Script

It becomes cumbersome to manually start and stop the master and workers as the number of workers increases in your Spark cluster. Fortunately, Spark provides scripts to start and stop the master and all the workers all at once.

To use these scripts, you need to first create a file called slaves in the /path/to/spark/conf directory. This file should list the hostnames or IP addresses of all the workers, one per line.

In addition, you need to enable private-key based *ssh* access from the master machine to all the worker machines. The launch script running on the master connects to all the workers in parallel using *ssh*.

After these requirements have been met, you can start a Spark cluster by running the start-all.sh script on the master node, as shown next.

```
$ /path/to/spark/sbin/start-all.sh
```

The start-all.sh script launches the master and workers with default configurations. You can specify custom configuration by setting environment variables in the spark-env.sh file in the /path/to/spark/conf directory. Spark provides a template file named spark-env.sh.template in the /path/to/spark/conf directory. This file lists all the configuration variables along with their descriptions. Make a copy of this file, rename it to spark-env.sh, and customize it as per your requirements. You need to do this on the master and all the worker nodes.

Stopping a Cluster with a Script

You can stop a cluster by running the stop-all.sh script on the master node, as shown next.

```
$ /path/to/spark/sbin/stop-all.sh
```

Running a Spark Application on a Standalone Cluster

A Spark application can be deployed on a standalone cluster using the spark-submit script. First, create a JAR file for your application. If it has dependencies on other external libraries, create an "uber" JAR that includes both your application code and all of its external dependencies.

A Spark application can be launched on a cluster, as shown next.

```
$ /path/to/spark/bin/spark-submit --master <master-URL> </path/to/app-jar> [app-arguments]
```

The spark-submit script is configurable; it accepts a number of command-line options. You can learn about all the options that can be passed to spark-submit using the help option.

```
$ /path/to/spark/bin/spark-submit --help
```

A Spark application can be deployed on a cluster in two modes: client and cluster. The spark-submit script provides a command-line option, --deploy-mode, which can used to run a Spark application in client or cluster mode. By default, it runs in the client mode.

■ **Note**   Remember that a Spark application is also referred to as the *driver program*. These terms are used interchangeably in this chapter.

Client Mode

In client mode, the driver program runs within the client process that is used to deploy a Spark application on a Spark cluster. For example, if you use the `spark-submit` script to deploy your Spark application, the `spark-submit` script is the client. In this case, the driver program, which is essentially your Spark application, runs within the same process that is running the `spark-submit` script.

In client mode, the standard input and standard output of the driver program is attached to the console from where it was launched. In addition, it can read from and write to local files on the machine from where it is launched.

It is recommended to use the client deployment mode only if the client machine is on the same network as the Spark cluster. You can use either the master node or another machine on the same network to launch a Spark application in client mode.

A Spark application can be launched in client mode, as shown next.

```
$ /path/to/spark/bin/spark-submit --deploy-mode client \
                        --master <master-URL> \
                        </path/to/app-jar> [app-arguments]
```

As mentioned previously, the default mode is the client mode, so you don't need to necessarily specify the `deploy-mode` if you want to run an application in client mode.

Cluster Mode

In cluster mode, the driver program runs on one of the worker nodes in a Spark cluster. The `spark-submit` script sends a request to the master to find a worker node for running the driver program. The driver program is then launched from one of the worker processes.

It is recommended to use the cluster mode if the machine from where you deploy a Spark application is not on the same network as the Spark cluster. This enables you to minimize network latency between the driver program and executors. This mode also gives you the option to have the driver program automatically restarted if it terminates with a non-zero exit code.

Note that when an application is deployed in cluster mode using the `spark-submit` script, the latter exits after submitting the application. In addition, since the driver program runs on one of the worker nodes, it does not have access to the console of the machine from where it was deployed. It needs to store its output in a file or a database.

A Spark application can be launched in cluster mode, as shown next.

```
$ /path/to/spark/bin/spark-submit --deploy-mode cluster \
                        --master <master-URL> \
                        </path/to/app-jar> [app-arguments]
```

When an application is deployed in cluster mode, the `spark-submit` script prints a submission ID, which can be used to check the status of the driver program or to stop it.

The status of a driver program running in cluster mode can be checked, as shown next.

```
$ /path/to/spark/bin/spark-submit --status <submission-id>
```

A driver program running in cluster mode can be stopped, as shown next.

```
$ /path/to/spark/bin/spark-submit --kill <submission-id>
```

If you want the driver program to be automatically restarted in case it terminates with a non-zero exit code, use the --supervise flag, as shown next.

```
$ /path/to/spark/bin/spark-submit --deploy-mode cluster --supervise \
                        --master <master-URL> \
                        </path/to/app-jar> [app-arguments]
```

Apache Mesos

Apache Mesos is an open source cluster manager. Conceptually, it can be thought of as an operating system kernel for a cluster of computers. It pools together the compute resources on a cluster of machines and enables sharing of those resources across diverse applications.

Mesos provides APIs in Java, C++, and Python for acquiring cluster resources and scheduling tasks across a cluster. In addition, it provides a web UI for monitoring a cluster.

One of the key benefits of Mesos is that it enables different distributed computing frameworks to dynamically share cluster resources. It enables both Spark and non-Spark applications to run simultaneously on the same cluster. For example, Mesos can run Spark, Hadoop, MPI, Kafka, Elastic Search, Storm, and other applications on a dynamically shared pool of nodes.

By enabling diverse computing frameworks to dynamically share a cluster, Mesos not only enables efficient utilization of cluster resources, but also prevents expensive data duplication. The same data can be processed by applications written with different frameworks without having to copy it from one cluster to another.

Mesos offers a lot more functionality than that provided by the Spark Standalone cluster manager. It not only enables different frameworks to share cluster resources, but also makes it easier to develop frameworks for distributed computing.

The following are the other key features of Mesos:

- Scalability to 10,000s of nodes

- Fault-tolerant masters and slaves

- Multi-resource (CPU, memory, disk, and ports) scheduling

- Support for Docker containers

- Native isolation between tasks

Architecture

The Mesos cluster manager consists of two key components: master and slave (see Figure 10-2). Similar to a worker in a standalone cluster, a Mesos slave manages the compute resources on a cluster node. The Mesos master manages the Mesos slaves and pools together the resources managed by each slave.

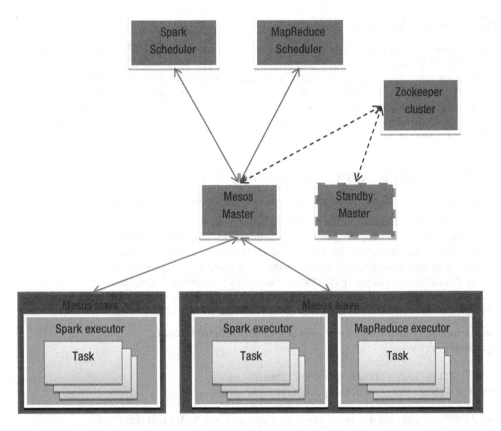

Figure 10-2. *Mesos architecture*

A framework or application implemented with Mesos consists of a scheduler and executors. The executors run on the slave nodes and execute application tasks.

To prevent single-point of failure, Mesos supports multiple masters. At any given point, only one master is active and others are in standby mode. If an active master fails, a new master is elected. Mesos uses Zookeeper to handle master failover.

When an application is deployed on a Mesos cluster, the scheduler component of the application connects to the Mesos master, which makes it a resource offer. A resource offer is a list of compute resources such as CPU cores, memory, disk, and port on the slave nodes. The amount of resources offered by the Mesos master to an application is configurable.

A scheduler can accept or reject an offer sent by the Mesos master. If it accepts an offer, the scheduler selects which of the offered resources to use. Next, it sends the application tasks to run on the offered resources to the Mesos master. The Mesos master forwards these tasks to the slaves, which allocate appropriate resources to the executors, which run the application tasks. This process repeats when the tasks finish and new resources become available.

Setting Up a Mesos Cluster

Mesos should be installed on each node that will be used as a Mesos master or slave. The instructions for downloading and installing Mesos are at http://mesos.apache.org/gettingstarted/.

You can run Mesos with a single master. In that case, you don't need to install Zookeeper. However, it is recommended to install Zookeeper to enable Mesos master failover. The instructions for installing Zookeeper are at http://zookeeper.apache.org/.

You can verify that the Mesos master and all the Mesos slaves are correctly set up by connecting to the Mesos master from a browser on port 5050.

Running a Spark Application on a Mesos Cluster

The following steps must be completed before deploying a Spark application on a Mesos cluster.

1. Make the Spark binaries accessible to all the Mesos slaves. When a Mesos slave runs a Spark task, it needs access to the Spark binaries for launching a Spark Executor. Place the Spark binaries at any Hadoop-accessible URI or the local file system on all the Mesos slave nodes. In addition, configure the variable spark. mesos.executor.home in the /path/to/spark/conf/spark-default.conf file to point to that location. By default, spark.mesos.executor.home points to the directory location specified by the environment variable SPARK_HOME.

2. Set the following environment variables in the /path/to/spark/conf/ spark-env.sh file.

    ```
    MESOS_NATIVE_JAVA_LIBRARY=/path/to/libmesos.so
    SPARK_EXECUTOR_URI=/path/to/spark-x.y.z.tar.gz
    ```

3. Set the following variables in the /path/to/spark/conf/spark-default.conf file.

    ```
    spark.executor.uri = /path/to/spark-x.y.z.tar.gz
    ```

You can launch a Spark application on a Mesos cluster using the spark-submit script. A Spark application can be launched on a single-master Mesos cluster, as shown next.

```
$ /path/to/spark/bin/spark-submit --master mesos://host:5050 </path/to/app-jar> [app-args]
```

Note that the master URL points to the Mesos master.
On a multi-master Mesos cluster, a Spark application can be launched, as shown next.

```
$ /path/to/spark/bin/spark-submit \
          --master mesos://zk://host1:2181,host2:2181,host3:2181 \
          </path/to/app-jar> [app-args]
```

In this case, the master URL points to the Zookeeper nodes.

Deploy Modes

Mesos supports both client and cluster mode deployment of an application. In client mode, the driver program runs within the client process on the machine from where it is deployed. In cluster mode, it runs on one of the Mesos slave nodes. However, at the time this book was being written, Spark does not support cluster mode for Mesos clusters.

Run Modes

A Spark application can run on a Mesos cluster in two modes: fine-grained and coarse-grained.

Fine-grained Mode

In fine-grained mode, each application task runs as a separate Mesos task. This mode enables dynamic sharing of cluster resources among different applications. Thus, it enables efficient utilization of cluster resources.

However, the fine-grained mode incurs higher overhead than the coarse-grained mode. It is not appropriate for iterative or interactive applications.

By default, a Spark application runs in fine-grained mode on a Mesos cluster.

Coarse-grained Mode

In coarse-grained mode, Spark launches a single long-running task on each Mesos slave for each Spark application. Jobs submitted by an application are split into tasks, which are run within this long-running task. In this mode, Spark reserves cluster resources for the complete duration of an application.

The coarse-grained mode incurs lower overhead but can result in inefficient utilization of cluster resources. For example, if you launch a Spark shell on a Mesos cluster, the cluster resources reserved for the Spark shell cannot used by any other application even if the Spark shell is idle.

To enable coarse-grained mode, set the following variables in the /path/to/spark/conf/spark-default.conf file.

```
spark.mesos.coarse=true
```

In coarse-grained mode, by default, Spark assign all the resources offered to it by the Mesos master to an application. Therefore, you cannot run multiple Spark applications simultaneously. You can change the default behavior by capping the resources Spark allocates to an application. The maximum number of CPU cores for an application can be configured in the /path/to/spark/conf/spark-default.conf file, as shown next.

```
spark.cores.max=N
```

Replace N with the right number as per your requirement.

YARN

YARN (yet another resource negotiator) is also a general open source cluster manager. It became available starting with Hadoop MapReduce version 2.0. In fact, Hadoop MapReduce version 2.0 is called YARN.

The first generation of Hadoop MapReduce supported only the MapReduce compute engine. It included a component called JobTracker, which provided cluster management and job scheduling as well as monitoring capabilities. The cluster manager and the compute engine were tightly coupled.

In Hadoop MapReduce version 2.0, cluster management and job scheduling/monitoring got split into separate processes. YARN provides cluster management capabilities and an application-specific master provides job scheduling and monitoring capabilities. With this new architecture, YARN can support not only the MapReduce compute engine, but also other compute engines such as Spark and Tez.

Architecture

The YARN cluster manager consists of two key components: ResourceManager and NodeManager (see Figure 10-3). The ResourceManager is the YARN equivalent of the Mesos master. The NodeManager is the YARN equivalent of the Mesos slave.

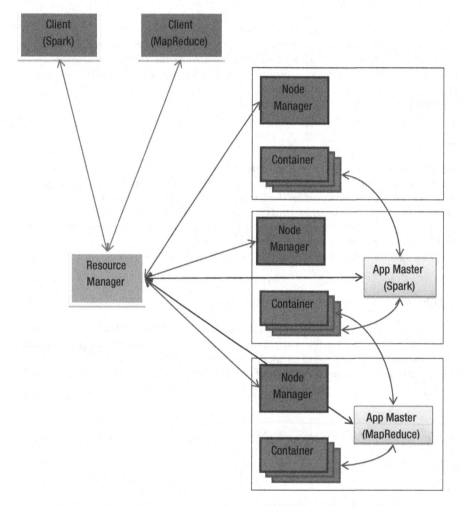

Figure 10-3. YARN architecture

The NodeManager manages the resources available on a single node. It reports these resources to the ResourceManager. The ResourceManager manages resources available across all the nodes in a cluster. It pools together the resources reported by all the NodeManagers and allocates them to different applications. Essentially, it is a scheduler that schedules available cluster resources among applications.

A YARN-based distributed computing framework consists of three key components: client application, ApplicationMaster, and containers.

A client application submits a job to the ResourceManager. For example, the spark-submit script is a client application.

The ApplicationMaster is generally provided by a library such as Spark or MapReduce. They create an ApplicationMaster per application. It owns and executes a job on a YARN cluster. The ApplicationMaster negotiates resources with the ResourceManager and works with the NodeManagers to execute a job using containers. It also monitors jobs and tracks progress. The ApplicationMaster runs on one of the machines running a NodeManager.

A container conceptually represents the resources available to an application on a single node. The ApplicationMaster negotiates the containers required to execute a job with the ResourceManager. On successful allocation, it launches containers on the cluster nodes working with the NodeManagers. The NodeManager manages the containers running on a machine.

The ResourceManager consists of two key components: ApplicationsManager and Scheduler. ApplicationsManager accepts jobs from a client application and assigns the first container for running the ApplicationMaster. The Scheduler allocates cluster resources to the ApplicationMaster for running jobs. It provides only scheduling functionality; it does not monitor or track application status.

One of the benefits of using YARN is that Spark and MapReduce applications can share the same cluster. If you have an existing Hadoop cluster, you can easily deploy Spark applications on it using YARN.

Running a Spark Application on a YARN Cluster

YARN also supports both client and cluster mode deployment of a Spark application. In client mode, the driver program runs in the client process that was used to deploy a Spark application. In this mode, the ApplicationMaster is used only for requesting resources from the ResourceManager. In cluster mode, the driver program runs within the ApplicationMaster process running on one of the cluster nodes.

To run a Spark application on a YARN cluster, the environment variable HADOOP_CONF_DIR or YARN_CONF_DIR must be configured in the /path/to/spark/conf/spark-env.sh file to point to the directory that contains client-side Hadoop configuration files.

The spark-submit script can be used to also deploy a Spark application on a YARN cluster. However, when launching a Spark application on a YARN cluster, the value for the master-URL is either yarn-cluster or yarn-client. Spark picks up the ResourceManager's address from the Hadoop configuration file.

A Spark application can be launched on YARN in cluster mode, as shown next.

```
$ /path/to/spark/bin/spark-submit --class path.to.main.Class \
    --master yarn-cluster </path/to/app-jar> [app-args]
```

A Spark application can be launched in client mode, as shown next.

```
$ /path/to/spark/bin/spark-submit --class path.to.main.Class \
    --master yarn-client </path/to/app-jar> [app-args]
```

Summary

With its modular architecture, Spark supports multiple cluster managers. A Spark application can be deployed on a Mesos, YARN, or standalone cluster.

The Standalone cluster manager is the simplest amongst the three cluster managers supported by Spark. It is easy to set up and get started. However, it supports only Spark applications and provides limited functionality.

YARN allows a Spark application to share cluster resources with Hadoop MapReduce applications. If you have an existing Hadoop cluster running MapReduce jobs, you can use YARN to run Spark job on the same cluster.

Mesos is the most general cluster manager amongst the three cluster managers covered in this chapter. It is designed to support a variety of distributed computing applications. It allows both static and dynamic sharing of cluster resources.

CHAPTER 11

Monitoring

Monitoring is a critical part of application management. It plays an even more important role in distributed computing, which involves a lot more moving parts. The possibility of something failing or application not performing at optimal level is high.

However, troubleshooting problems and optimizing application performance are difficult tasks in a distributed environment. Unlike an application running on a single machine, a distributed system cannot be debugged with traditional debugging tools. Therefore, it is essential to instrument the various components of a distributed system, so that they can be remotely monitored and analyzed.

In addition to instrumenting a distributed system, a monitoring application, which collects and displays metrics, is required. A monitoring application plays a similar role as an instrument cluster in a car dashboard or flight instruments in an airplane cockpit. It lets you track important metrics and allows you to view the status of different components in real time.

Spark provides strong monitoring capabilities. The various components of Spark are heavily instrumented. In addition, it comes pre-packaged with a web-based monitoring application, which can be used to monitor both the Standalone cluster manager and Spark applications. It also supports third-party monitoring tools such as Ganglia, Graphite, and JMX-based monitoring tools.

This chapter describes the built-in web-based monitoring application that come with Spark.

Monitoring a Standalone Cluster

Spark provides a web-based UI for monitoring both the master and the workers in a Spark standalone cluster. You can access the built-in monitoring UI by connecting to a Spark master or a worker on its monitoring port from a browser.

Monitoring a Spark Master

The default TCP port for the monitoring UI for a Spark master is 8080. The monitoring port is configurable; if port 8080 is not available, you can configure a Spark master to use another port.

An example of the web UI for monitoring a Spark master is shown in Figure 11-1.

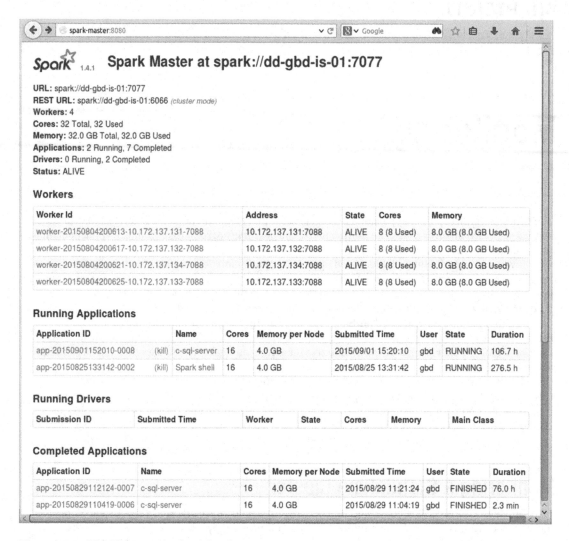

Figure 11-1. Web UI for monitoring a Spark master

The monitoring page consists of multiple sections. At the top, it shows the Spark version and the URL for the Spark master. This is the URL that a worker needs to use to connect to this master.

The next section shows summary information about a cluster. It shows the number of workers in a cluster, number of cores available and in use across all the worker nodes, total memory available and in use across all the worker nodes, number of applications currently running and completed, and the current status of the master.

The summary section is followed by the Workers section, which provides summary information about all the worker nodes known to the master. For each worker, this section shows its location, current state, the number of cores available and in use, and the memory available and in use.

The Workers section is followed by the Running Applications section. It shows summary information about the currently running applications. For each application, this section shows its name, current state, resources allocated, start time, owner, and the amount of time that it has been running.

Note that an application may be granted different number of cores on different workers. The Cores column shows the sum of the cores allocated on different worker nodes to an application.

The section titled Completed Applications shows information about the Spark applications that are no longer running on the cluster. It has similar columns as the Running Applications section. The state column shows whether an application finished normally or stopped because of some error.

The web UI for the Spark master is useful for troubleshooting cluster-related problems. For example, if a worker node crashes or did not start correctly, its state will be shown as DEAD. Similarly, the cores and memory-related metrics for each worker node can help you identify configuration and resource allocation as well as usage-related issues.

The cores and memory available to a worker should match with the configuration settings for that worker. If it does not, it indicates that either the configurations variables are not correctly initialized or Spark is picking up a different configuration file.

The cores and memory in use on the worker nodes are useful metrics for identifying and troubleshooting potential performance issues. Ideally, the cores and memory in use should be same on all the worker nodes. If one worker shows 100% of the available cores and memory in use and another worker shows only 25% of the available cores and memory in use, it indicates that the cluster resources are not evenly allocated. As a result, applications may not perform optimally. You may need to stop all the applications and restart them again.

The cores and memory-related metrics are also useful for determining whether cluster resources are available for launching a new application on a Spark cluster. If the total and used number is same for either memory or cores, it means that the Spark scheduler cannot assign a core or memory to a new application. For example, for the cluster shown in Figure 11-1, all the available cores and memory have been allocated to the applications currently running on the cluster. The used number is same as the total (available) for both cores and memory. Until one of the existing application finishes, jobs from a new application cannot be executed on this cluster.

When you launch a Spark application on a cluster, check the state of the application under the Running Applications section. If it is in WAITING state, it indicates that Spark does not have either the required cores or memory for the application. It will remain in that state until resources become available. You can wait until one of the existing applications finishes, increase the resources allocated to the Spark workers, or reduce the resources requested by the new application.

The Completed Applications section is useful for both identifying performance issues and troubleshooting applications that did not finish normally. The State column shows whether an application finished normally or failed. For failed applications, you can check the logs for more details. The duration column shows how long an application took to finish its jobs. Depending on the amount of the data processed, the duration will vary from application to application. You can determine from the duration whether there are performance issues.

The monitoring page for the master also serves as a portal for the workers' and applications' monitoring UI. The Workers section provides links to the monitoring UI for each worker node. Similarly, the Applications section provides links to the monitoring UI for each Spark application deployed on the cluster.

When you click the application id of an application, you get the page shown in Figure 11-2. It shows summary information about an application and its executors.

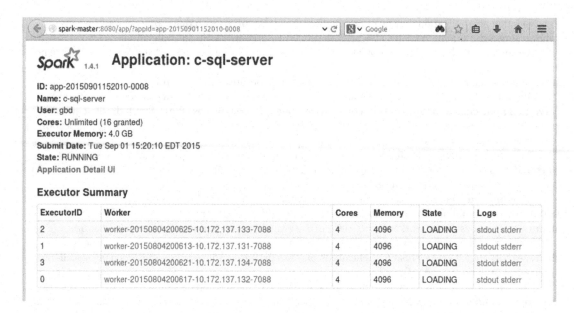

Figure 11-2. *Web UI for monitoring executors*

The top section shows application name, owner, cores requested and allocated, memory allocated to each executor, launch time, and current state.

The Executor Summary section provides a summary of each executor launched for an application. The executors execute application tasks. For each executor, this section shows the worker node on which it is running, resources allocated to it, its current state, and the links to its logs files.

This page provides a quick view of the resources allocated to an application on each worker node. Ideally, the cores and memory allocated should be same on each node. Uneven resource allocation can potentially cause performance issues.

This page also allows you to remotely view the executor logs. The log files contain useful information for troubleshooting problems. You can also browse the log files to get a better understanding of how Spark works.

A Spark executor writes all logs, including DEBUG and INFO logs, to the stderr file. The stdout file is empty. If a task fails or throws an exception, check the stderr file to debug the problem. You can also use the log files to analyze performance issues.

Monitoring a Spark Worker

Similar to the Spark master, Spark workers can be remotely monitored from a browser. The default TCP port for accessing the monitoring web UI for a worker is 8081. If port 8081 is unavailable, you can configure a Spark worker to use another port.

An example of the web UI for monitoring a Spark worker is shown in Figure 11-3.

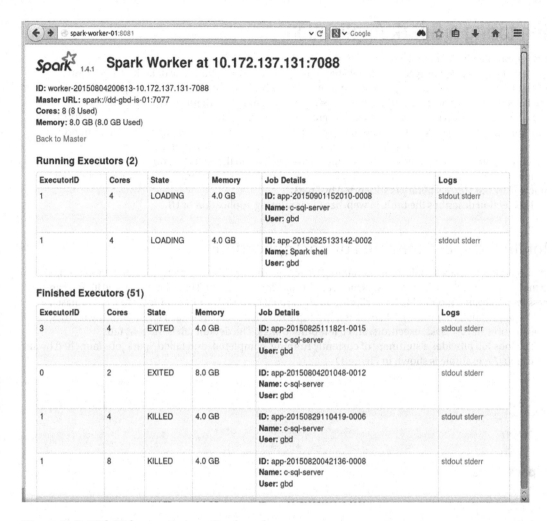

Figure 11-3. Web UI for monitoring a Spark worker

The worker monitoring UI shows summary information about a Spark worker and the executors launched on that node.

The top part of the page shows the Spark version and the location of a worker. Next, it shows the master to which a worker is connected to and the resources available and used on the worker.

The next section shows the executors currently running on a worker node. For each executor, this section shows its current state, resources allocated, and links to the log files. It also shows the owner and the application for which an executor was launched.

The last section shows the same information about completed executors. The only difference is that the state column shows how an executor terminated.

The worker monitoring page is useful for checking how the resources on a worker node are allocated to different applications. Depending on your requirements, you may evenly distribute the resources among different applications or assign different amount to different application.

You can also use the worker monitoring page to remotely check the log files. These are the same log files that were discussed in the previous section. The difference is that the worker monitoring page allows you to check the logs files generated by different executors running on a worker, whereas the executor monitoring page allows you to check the logs files generated by the executors running on different workers.

247

Monitoring a Spark Application

Spark supports a variety of tools for monitoring a Spark application.

First, it comes prepackaged with a web-based monitoring application. The default TCP port for accessing the built-in monitoring application from a browser is 4040. If multiple Spark applications are launched on the same host, they bind to successive ports beginning with port 4040.

Second, Spark provides a REST API for programmatically accessing monitoring metrics. The REST API returns the metrics in JSON format. This allows an application developer to create a custom monitoring application or integrate Spark application monitoring with other application monitoring.

Third, Spark can be configured to send application metrics to third-party monitoring tools such as Graphite, Ganglia, and JMX-based consoles. Spark uses the Metrics Java library, so any monitoring tool supported by the Metrics library is supported by Spark.

This section describes the built-in web-based monitoring application or UI.

Monitoring Jobs Launched by an Application

You can connect to Spark's application monitoring UI from a browser using the driver program's monitoring URL. Use the hostname or IP address of the machine running the driver program. The default port is 4040. You can also access the application monitoring UI by clicking an application name in the monitoring UI for a Spark master.

The application monitoring UI consists of multiple tabs. It has separate tabs for monitoring metrics related to jobs, stages, tasks, executors, and a few other things. The default tab is the Jobs tab.

The Jobs tab provides a summary of currently running, completed, and failed Spark jobs launched by an application. An example is shown in Figure 11-4.

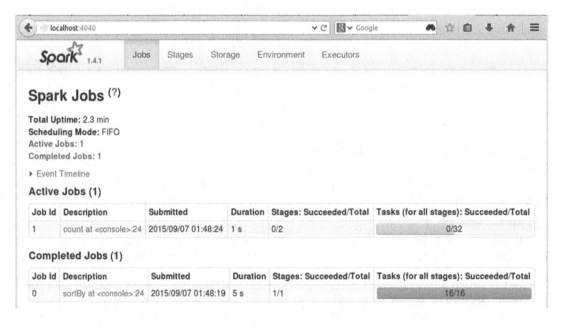

Figure 11-4. *Web UI for monitoring a Spark application*

The top part of the page shows application uptime and scheduling mode. It also shows the total number of active, completed, and failed jobs.

The Active Jobs section provides a summary of the active jobs. It shows the number of total and completed stages and tasks for each job. In addition, it shows the description, launch time, and duration of each job.

The Completed Jobs section provides a summary of the completed jobs. It displays similar information as the Active Jobs section.

The last section provides a summary of the failed jobs. It displays similar information as the preceding two sections for jobs that did not complete.

The Jobs tab is useful for detecting performance issues. The duration column shows the time taken by each job. Depending on the amount of data processed, you can determine whether a job is taking or took abnormally long time to complete. If a job takes abnormally long time, you can drill down to analyze the problematic stage or task.

Monitoring Stages in a Job

The job descriptions on the Jobs tab are hyperlinks. When you click a job description, it takes you to a page that shows more details about a job.

An example of the web UI for viewing job details is shown in Figure 11-5.

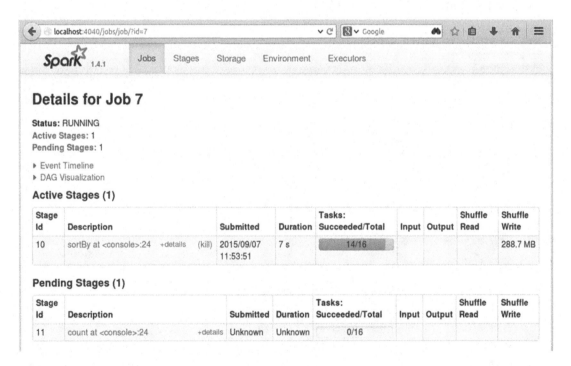

Figure 11-5. *Web UI for monitoring job details*

The Details page is split into multiple sections. The first section shows the total number of active and pending stages in a job as well as the current status of a job.

The Active Stages section shows a summary of the currently running stages. The next section shows summary information for the stages that are waiting for execution. A stage in pending state has a dependency on one or more active stages.

The information provided by the different columns can be useful for troubleshooting performance issues. For example, you can check the Duration column to find the stages that are taking abnormally long time. The Tasks column shows the amount of parallelism within a stage. Depending on the size of your cluster, too few or too many tasks can be a source of a performance issue. Similarly, the information provided by the Shuffle Read and Write columns can be used to optimize application performance. Data shuffle negatively impacts the performance of an application, so minimize the amount of shuffle reads and writes.

You can also kill a job from this page. If a job is taking too long and blocking other jobs or if you want to make code changes and run a long running job again, you can kill it from here. The description column provides a link to kill a job.

The Details page also includes links that show event timeline of a job and visual representation of the DAG generated by Spark for a job.

A sample visual representation of a DAG is shown in Figure 11-6.

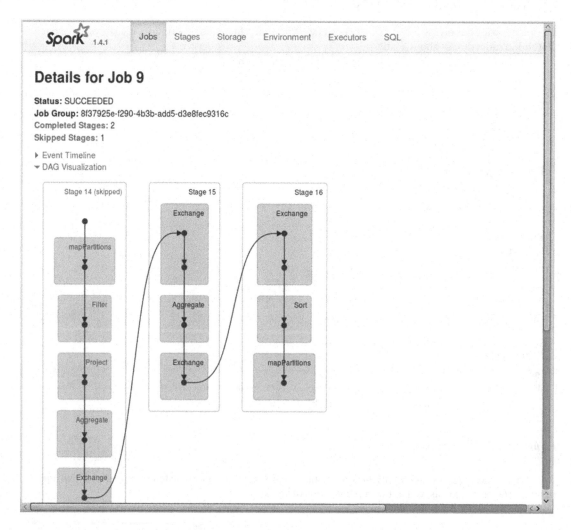

Figure 11-6. *Visual representation of a DAG*

Monitoring Tasks in a Stage

When you click a stage's description shown on the Jobs page, the web UI takes you to a page that shows detailed information about the tasks in a stage.

This is one of the most useful pages for troubleshooting performance issues. It is an information-rich page filled with a lot of useful information. It provides more fine-grained information compared to the other monitoring pages discussed so far.

Take time to familiarize yourself with the information provided on this page. It will help you understand better how jobs are executed on a Spark cluster.

An example web UI page that shows the details for a stage is shown in Figure 11-7.

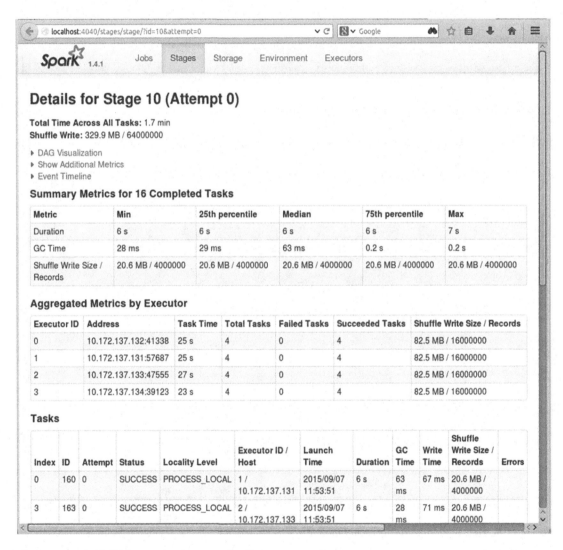

Figure 11-7. *Web UI for monitoring stage details*

Similar to the other monitoring UI pages, the stage details page is split into multiple sections. It begins with summary information about a stage. Next, it provides links for viewing the DAG for a stage, additional metrics for a stage, and timeline for tasks in a stage.

When you click the event timeline link, it shows the timeline for the tasks in a stage. It shows not only how many tasks are running in parallel on each worker node, but also what activities contribute to the total time taken by a task to complete.

An example of the timeline graph is shown in Figure 11-8. In this example, each executor runs four tasks in parallel. The tasks are compute-intensive with little data shuffle.

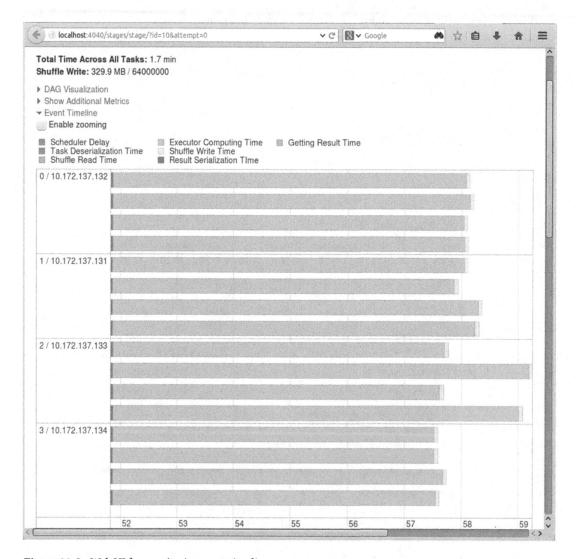

Figure 11-8. *Web UI for monitoring event timeline*

Figure 11-9 shows another example. It shows the event timeline graph for a stage with different characteristics.

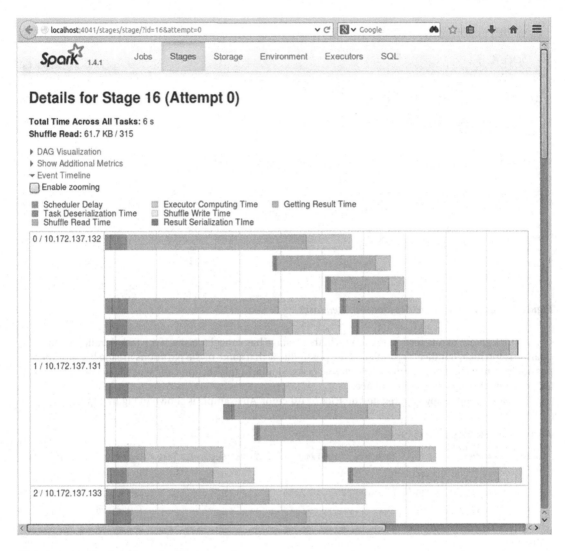

Figure 11-9. Another event timeline example

The next section shows a summary of the metrics for the completed tasks in a stage. An example is shown in Figure 11-10.

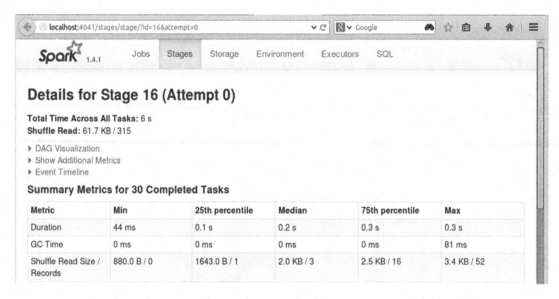

Figure 11-10. Summary metrics for completed tasks

A quick glance at this section tells you whether a stage has straggler tasks. If the task duration in any quartile is too high, it indicates problem. One possibility is that some of the partitions are too big. Another possibility is that shuffle is negatively impacting tasks. Similarly, this section can be used to check whether GC activities are impacting performance.

The next section shows aggregated metrics by executor. An example is shown in Figure 11-11.

Aggregated Metrics by Executor

Executor ID	Address	Task Time	Total Tasks	Failed Tasks	Succeeded Tasks	Shuffle Read Size / Records
0	10.172.137.132:44467	2 s	9	0	9	13.4 KB / 65
1	10.172.137.131:44615	2 s	8	0	8	16.3 KB / 100
2	10.172.137.133:37458	1 s	5	0	5	11.1 KB / 79
3	10.172.137.134:33189	2 s	8	0	8	20.9 KB / 71

Figure 11-11. Aggregated metrics by executor

The last section shows information about each task in a stage. It shows the key metrics for each task. An example is shown in Figure 11-12.

localhost:4041/stages/stage/?id=16&attempt=0

Tasks

Index	ID	Attempt	Status	Locality Level	Executor ID / Host	Launch Time	Duration	GC Time	Shuffle Read Size / Records	Errors
0	851	0	SUCCESS	PROCESS_LOCAL	2 / 10.172.137.133	2015/09/07 12:04:33	0.2 s		2.1 KB / 16	
4	855	0	SUCCESS	PROCESS_LOCAL	2 / 10.172.137.133	2015/09/07 12:04:33	0.3 s		3.4 KB / 47	
6	857	0	SUCCESS	PROCESS_LOCAL	3 / 10.172.137.134	2015/09/07 12:04:33	0.3 s		2.9 KB / 14	
9	860	0	SUCCESS	PROCESS_LOCAL	1 / 10.172.137.131	2015/09/07 12:04:33	0.1 s		2.6 KB / 16	
3	854	0	SUCCESS	PROCESS_LOCAL	0 / 10.172.137.132	2015/09/07 12:04:33	0.3 s		1881.0 B / 31	
11	862	0	SUCCESS	PROCESS_LOCAL	0 / 10.172.137.132	2015/09/07 12:04:33	0.2 s		1846.0 B / 5	
2	853	0	SUCCESS	PROCESS_LOCAL	3 / 10.172.137.134	2015/09/07 12:04:33	0.3 s		3.4 KB / 30	
5	856	0	SUCCESS	PROCESS_LOCAL	1 / 10.172.137.131	2015/09/07 12:04:33	0.3 s		2.3 KB / 23	
8	859	0	SUCCESS	PROCESS_LOCAL	2 / 10.172.137.133	2015/09/07 12:04:33	0.3 s		2.4 KB / 14	
7	858	0	SUCCESS	PROCESS_LOCAL	0 / 10.172.137.132	2015/09/07 12:04:33	0.3 s		1050.0 B / 17	
10	861	0	SUCCESS	PROCESS_LOCAL	3 / 10.172.137.134	2015/09/07 12:04:33	0.3 s		3.0 KB / 18	
1	852	0	SUCCESS	PROCESS_LOCAL	1 / 10.172.137.131	2015/09/07 12:04:33	0.2 s		2.5 KB / 52	
13	864	0	SUCCESS	PROCESS_LOCAL	3 / 10.172.137.134	2015/09/07 12:04:33	0.3 s		2.6 KB / 4	
12	863	0	SUCCESS	PROCESS_LOCAL	1 / 10.172.137.131	2015/09/07 12:04:33	0.2 s		1306.0 B / 2	
14	865	0	SUCCESS	PROCESS_LOCAL	0 / 10.172.137.132	2015/09/07 12:04:33	0.2 s		1408.0 B / 8	

Figure 11-12. Task metrics

This section is useful for identifying problematic tasks. It shows the duration of each task and the amount of data processed by it. You can use these two columns to find slow or straggler tasks. In addition, you should inspect the GC time to check whether tasks are running slow because of GC activities.

Another useful metric is the data locality level. The locality level PROCESS_LOCAL indicates that the data processed by a task is cached in memory. The locality level NODE_LOCAL means data was read from local storage and ANY indicates the data could have come from any node in a cluster. Tasks processing data with locality level PROCESS_LOCAL will be extremely fast.

You can also use this table to determine how many tasks are running in parallel both within each executor and across different executors. Thus, it gives you a picture of parallelism both across nodes and within an executor.

Figure 11-13 shows another example. It shows the Details page for another stage.

Details for Stage 234

Total task time across all tasks: 13 min
Input: 1168.1 MB

▸ **Show additional metrics**

Summary Metrics for 36 Completed Tasks

Metric	Min	25th percentile	Median	75th percentile	Max
Duration	5 s	16 s	22 s	27 s	30 s
GC Time	0 ms	0.4 s	0.5 s	0.6 s	0.6 s
Input	4.0 MB	28.8 MB	33.8 MB	37.7 MB	51.6 MB

Aggregated Metrics by Executor

Executor ID	Address	Task Time	Total Tasks	Failed Tasks	Succeeded Tasks	Input	Output	Shuffle Read	Shuffle Write	Shuffle Spill (Memory)	Shuffle Spill (Disk)
0	10.0.10.215:37733	4.2 min	13	0	13	426.0 MB	0.0 B	0.0 B	0.0 B	0.0 B	0.0 B
1	10.0.10.214:48268	6.1 min	15	0	15	467.7 MB	0.0 B	0.0 B	0.0 B	0.0 B	0.0 B
2	10.0.10.213:51753	2.3 min	8	0	8	274.5 MB	0.0 B	0.0 B	0.0 B	0.0 B	0.0 B

Tasks

Index	ID	Attempt	Status	Locality Level	Executor ID / Host	Launch Time	Duration	GC Time	Input	Errors
12	19710	0	SUCCESS	NODE_LOCAL	1 / 10.0.10.214	2015/12/04 06:56:59	27 s	0.5 s	44.4 MB (hadoop)	
3	19712	0	SUCCESS	NODE_LOCAL	2 / 10.0.10.213	2015/12/04 06:56:59	21 s	0.2 s	36.5 MB (hadoop)	
6	19711	0	SUCCESS	NODE_LOCAL	0 / 10.0.10.215	2015/12/04 06:56:59	27 s	0.6 s	51.6 MB (hadoop)	
18	19713	0	SUCCESS	NODE_LOCAL	1 / 10.0.10.214	2015/12/04 06:56:59	25 s	0.5 s	32.6 MB (hadoop)	
10	19715	0	SUCCESS	NODE_LOCAL	2 / 10.0.10.213	2015/12/04 06:56:59	14 s	0.2 s	39.6 MB (hadoop)	
14	19714	0	SUCCESS	NODE_LOCAL	0 /	2015/12/04	16 s	0.6 s	43.8 MB	

Figure 11-13. *Details for a stage*

Monitoring RDD Storage

The amount of data cached by a Spark application in memory or disk can be viewed by clicking the Storage tab in the web UI. This page provides a summary of each persisted RDD.

The Storage Level column shows how a dataset was cached. In addition, it shows the number of replicas of the cached data. The Size in Memory column is a useful column for troubleshooting. Deserialized data can take up a lot more memory space than serialized data.

The example in Figure 11-14 shows the Storage tab for an application that cached a Hive table in memory.

Figure 11-14. *Cached Hive table*

The example in Figure 11-15 shows the Storage tab for application that cached an RDD in memory.

Figure 11-15. *Cached RDD*

The names shown in the RDD Name column are hyperlinks. When you click an entry in this column, the web UI takes you to a page with more detailed information about the persisted RDD.

An example page with detailed RDD storage information is shown in Figure 11-16.

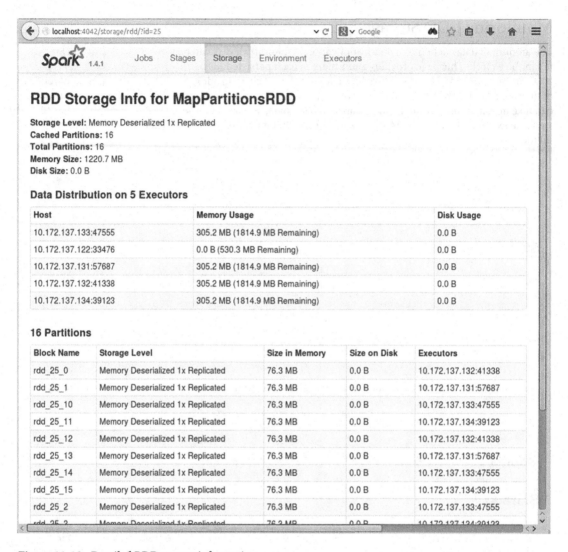

Figure 11-16. *Detailed RDD storage information*

The first section shows summary information about a cached RDD. The information shown here is same as shown in the previous figure.

The second section shows the distribution of a cached RDD across different executors. It shows the memory used on each executor for caching an RDD.

The third section shows information about each cached RDD partition. It shows the storage level, location, and size of each cached RDD partition.

Monitoring Environment

The Environment tab displays the values for the different environment and configuration variables. A sample page is shown in Figure 11-17.

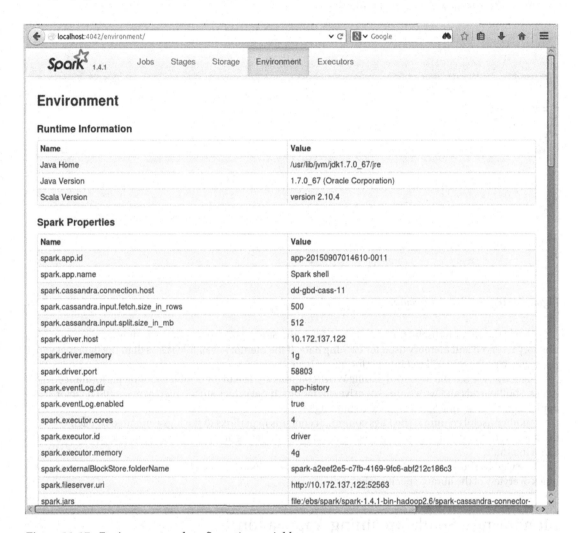

Figure 11-17. *Environment and configuration variables*

The Environment tab is useful for verifying whether the configuration and environment variables are correctly set. Reviewing them can help you identify potential problems. For example, some of the variables can be configured in multiple ways. You can set them programmatically, pass them as command-line arguments, or specify them in a configuration file. Spark defines the precedence order in such cases. However, if you forget the precedence order, you may end up running a Spark application with a configuration different from the one that you intended.

In addition, check the default values for the variables that you did not set explicitly. If the default values are not right for your applications, configure those variables explicitly.

Monitoring Executors

The Executors tab provides summary information about the executors that Spark creates for an application. At a glance, you can see useful metrics for each executor executing application tasks.

An example is shown in Figure 11-18.

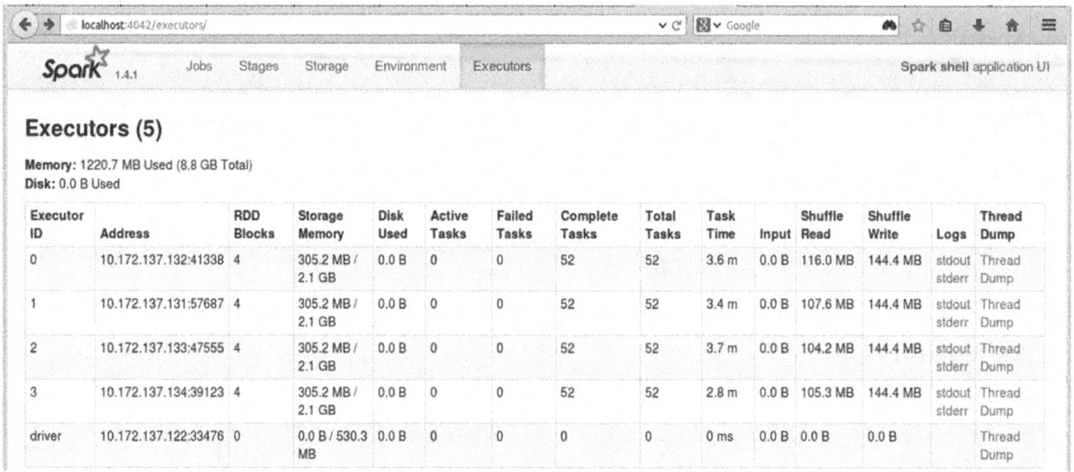

Figure 11-18. *Monitoring executors*

One of the columns useful for troubleshooting is the Storage Memory column. It shows the amount of memory reserved and memory used for caching data. If the memory available is less than the data that you are trying to cache, you will run into performance issues.

Similarly, the Shuffle Read and Shuffle Write columns are useful for troubleshooting performance issues. Shuffle reads and writes are expensive operations. If the values are too high, you should refactor application code or tune Spark to reduce shuffling.

Another useful column is the Logs column. It contains hyperlinks to the logs generated by each executor. As discussed earlier, you can use them to not only debug an application but also learn more about Spark internals.

The web UI shows a few additional tabs for some of the Spark libraries. The next few sections provide a quick overview of the library-specific tabs.

Monitoring a Spark Streaming Application

The web UI shows a tab for visualizing the execution of a Spark Streaming application if you are running one. The Streaming tab shows metrics that are useful for troubleshooting a Spark Streaming application.

An example is shown in Figure 11-19.

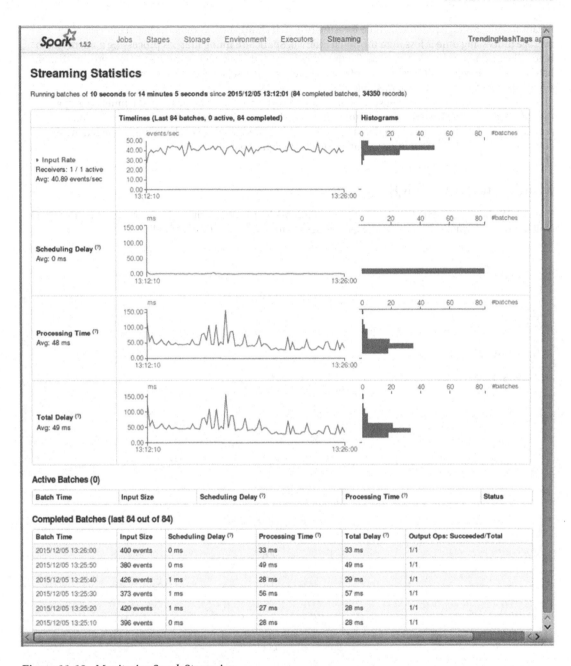

Figure 11-19. Monitoring Spark Streaming

The Streaming tab shows an information-rich page with many useful metrics. It shows the key indicators for detecting performance issues with a Spark Streaming application.

The metrics that you should specifically pay attention to are the average scheduling delay and average processing time for the micro-batches in a data stream. Scheduling delay is the time that a micro-batch waits for the previous micro-batch to complete. The average scheduling delay should be almost zero. This means that the average processing time for a micro-batch should be less than the batch interval. If the average processing time is less than the batch interval, Spark Streaming finishes processing a micro-batch before the next micro-batch is created. On the other hand, if the average processing time is greater than the batch interval, it will result in a backlog of micro-batches. The backlog will grow over a period of time and eventually make the application unstable.

The Streaming tab makes it easy to analyze the straggler batches. The peaks in the timeline graph that shows processing time represent a jump in processing time. If you click a peak, it will take you to the corresponding batches in the Completed Batches section. There you can click the link in the Batch Time column to see detailed information about a batch with high processing time.

Monitoring Spark SQL Queries

The web UI shows the Spark SQL tab if you execute Spark SQL queries. Figure 11-20 shows a sample SQL tab, which shows information about Spark SQL queries submitted from the Spark Shell.

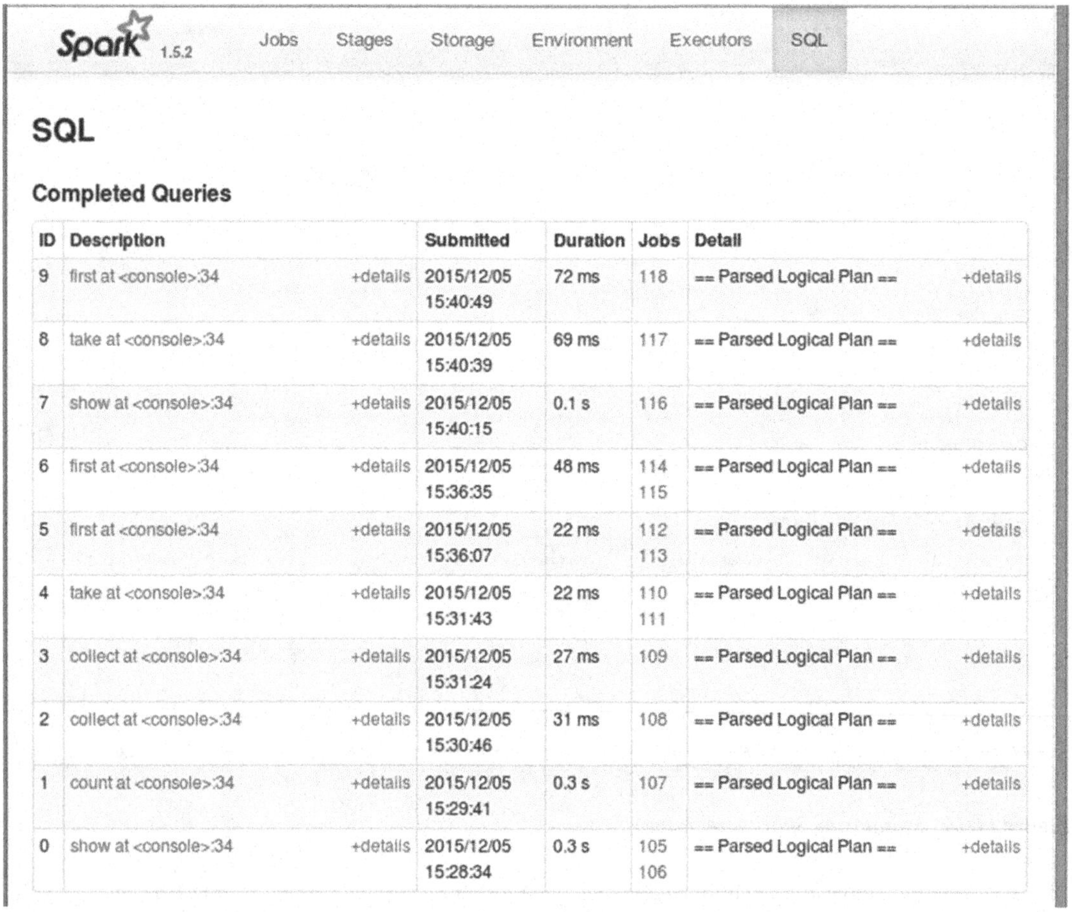

ID	Description		Submitted	Duration	Jobs	Detail	
9	first at <console>:34	+details	2015/12/05 15:40:49	72 ms	118	== Parsed Logical Plan ==	+details
8	take at <console>:34	+details	2015/12/05 15:40:39	69 ms	117	== Parsed Logical Plan ==	+details
7	show at <console>:34	+details	2015/12/05 15:40:15	0.1 s	116	== Parsed Logical Plan ==	+details
6	first at <console>:34	+details	2015/12/05 15:36:35	48 ms	114 115	== Parsed Logical Plan ==	+details
5	first at <console>:34	+details	2015/12/05 15:36:07	22 ms	112 113	== Parsed Logical Plan ==	+details
4	take at <console>:34	+details	2015/12/05 15:31:43	22 ms	110 111	== Parsed Logical Plan ==	+details
3	collect at <console>:34	+details	2015/12/05 15:31:24	27 ms	109	== Parsed Logical Plan ==	+details
2	collect at <console>:34	+details	2015/12/05 15:30:46	31 ms	108	== Parsed Logical Plan ==	+details
1	count at <console>:34	+details	2015/12/05 15:29:41	0.3 s	107	== Parsed Logical Plan ==	+details
0	show at <console>:34	+details	2015/12/05 15:28:34	0.3 s	105 106	== Parsed Logical Plan ==	+details

Figure 11-20. *Monitoring Spark SQL queries*

The SQL tab makes it easy to troubleshoot Spark SQL queries. The Detail column provides a link to the logical and physical plan generated by Spark SQL for a query. If you click the details link, it shows you the parsed, analyzed, and optimized logical plan, and the physical plan for a query.

The Jobs column is also useful for analyzing slow running queries. It shows the Spark jobs created by Spark SQL for executing a query. If a query takes a long time to complete, you can analyze it by clicking the corresponding job id shown in the Jobs column. It takes you to the page that shows the stages in a job. You can further drill down from there to the individual tasks.

Monitoring Spark SQL JDBC/ODBC Server

The monitoring web UI shows the JDBC/ODBC Server tab if the Spark SQL JDBC/ODBC server is running. The JDBC/ODBC Server tab allows you to monitor SQL queries submitted to the Spark SQL JDBC/ODBC server.

An example is shown in Figure 11-21.

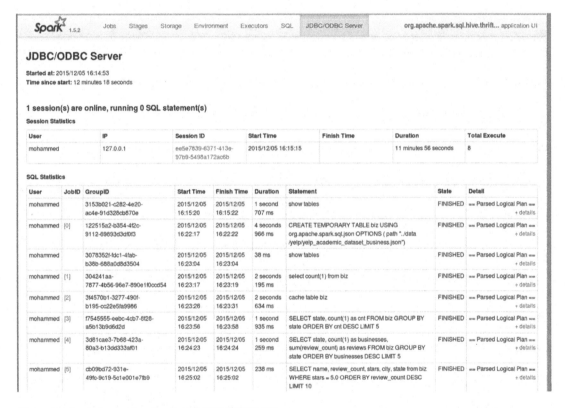

Figure 11-21. *Monitoring Spark SQL JDBC/ODBC Server*

Similar to the SQL tab, the JDBC/ODBC Server tab makes it easy to troubleshoot Spark SQL queries. It shows you the SQL statements and the time they took to complete. You can analyze a slow running query by checking its logical and physical plan. It also allows you to drill down into the Spark jobs created for executing queries.

Summary

Spark exposes a wealth of monitoring metrics and it comes pre-packaged with web-based applications that can be used to monitor a Spark standalone cluster and Spark applications. In addition, it supports third-party monitoring tools such as Graphite, Ganglia, and JMX-based consoles.

The monitoring capabilities provided by Spark are useful for both troubleshooting and optimizing application performance. The monitoring web UI helps you find configuration issues and performance bottlenecks. If a job is taking too long, you can use the web UI to analyze and troubleshoot it. Similarly, if an application crashes, the monitoring UI allows you to remotely check the log files and diagnose the problem.

Bibliography

Armbrust, Michael, Reynold S. Xin, Cheng Lian, Yin Huai, Davies Liu, Joseph K. Bradley, Xiangrui Meng, Tomer Kaftan, Michael J. Franklin, Ali Ghodsi, Matei Zaharia. *Spark SQL: Relational Data Processing in Spark.* https://amplab.cs.berkeley.edu/publication/spark-sql-relational-data-processing-in-spark.

Avro. http://avro.apache.org/docs/current.

Ben-Hur, Asa and Jason Weston. *A User's Guide to Support Vector Machines.* http://pyml.sourceforge.net/doc/howto.pdf.

Breiman, Leo. Random Forests. https://www.stat.berkeley.edu/~breiman/randomforest2001.pdf.

Cassandra. http://cassandra.apache.org.

Chang, Fay, Jeffrey Dean, Sanjay Ghemawat, Wilson C. Hsieh, Deborah A. Wallach, Mike Burrows, Tushar Chandra, Andrew Fikes, and Robert E. Gruber. Bigtable: A Distributed Storage System for Structured Data. http://research.google.com/archive/bigtable.html.

Dean, Jeffrey and Sanjay Ghemawat. *MapReduce Tutorial.* http://hadoop.apache.org/docs/current/hadoop-mapreduce-client/hadoop-mapreduce-client-core/MapReduceTutorial.html.

Drill. https://drill.apache.org/docs.

Hadoop. http://hadoop.apache.org/docs/current.

Hastie, Trevor, Robert Tibshirani, Jerome Friedman. *The Elements of Statistical Learning.* http://statweb.stanford.edu/~tibs/ElemStatLearn.

HBase. http://hbase.apache.org.

HDFS Users Guide. http://hadoop.apache.org/docs/current/hadoop-project-dist/hadoop-hdfs/HdfsUserGuide.html.

He, Yongqiang, Rubao Lee, Yin Huai, Zheng Shao, Namit Jain, Xiaodong Zhang, Zhiwei Xu. *RCFile: A Fast and Space-efficient Data Placement Structure in MapReduce-based Warehouse Systems.* http://web.cse.ohio-state.edu/hpcs/WWW/HTML/publications/papers/TR-11-4.pdf.

Hinton, Geoffrey. *Neural Networks for Machine Learning.* https://www.coursera.org/course/neuralnets.

Hive. http://hive.apache.org.

Impala. http://impala.io.

James, Gareth, Daniela Witten, Trevor Hastie, and Robert Tibshirani. *An Introduction to Statistical Learning.* http://www-bcf.usc.edu/~gareth/ISL/index.html.

Kafka. http://kafka.apache.org/documentation.html.

Koller, Daphne. *Probabilistic Graphical Models*. `https://www.coursera.org/course/pgm`.

Malewicz, Grzegorz, Matthew H. Austern, Aart J. C. Bik, James C. Dehnert, Ilan Horn, Naty Leiser, and Grzegorz Czajkowski. *Pregel: A System for Large-Scale Graph Processing*. `http://dl.acm.org/citation.cfm?doid=1807167.1807184`.

MapReduce: Simplified Data Processing on Large Clusters. `http://research.google.com/archive/mapreduce.html`.

Meng, Xiangrui, Joseph Bradley, Burak Yavuz, Evan Sparks, Shivaram Venkataraman, Davies Liu, Jeremy Freeman, DB Tsai, Manish Amde, Sean Owen, Doris Xin, Reynold Xin, Michael J. Franklin, Reza Zadeh, Matei Zaharia, Ameet Talwalkar. *MLlib: Machine Learning in Apache Spark*. `http://arxiv.org/abs/1505.06807`.

Mesos. `http://mesos.apache.org/documentation/latest`.

Ng, Andrew. *Machine Learning*. `https://www.coursera.org/learn/machine-learning`.

Odersky, Martin, Lex Spoon, and Bill Venners. *Programming in Scala*. `http://www.artima.com/shop/programming_in_scala_2ed`.

ORC. `https://orc.apache.org/docs`.

Parquet. `https://parquet.apache.org/documentation/latest`.

Presto. `https://prestodb.io`.

Protocol Buffers. `https://developers.google.com/protocol-buffers`.

Sanjay Ghemawat, Howard Gobioff, and Shun-Tak Leung. *The Google File System*. `http://research.google.com/archive/gfs.html`.

Scala. `http://www.scala-lang.org/documentation/`.

Sequence File. `https://wiki.apache.org/hadoop/SequenceFile`.

Sergey Melnik, Andrey Gubarev, Jing Jing Long, Geoffrey Romer, Shiva Shivakumar, Matt Tolton, Theo Vassilakis. *Dremel: Interactive Analysis of WebScale Datasets*. `http://research.google.com/pubs/archive/36632.pdf`.

Spark API. `http://spark.apache.org/docs/latest/api/scala/index.html`.

Thrift. `https://thrift.apache.org`.

Thrun, Sebastian and Katie Malone. *Intro to Machine Learning*. `https://www.udacity.com/course/intro-to-machine-learning--ud120`.

Xin, Reynold S., Daniel Crankshaw, Ankur Dave, Joseph E. Gonzalez, Michael J. Franklin, Ion Stoica. *GraphX: Unifying Data-Parallel and Graph-Parallel Analytics*. `https://amplab.cs.berkeley.edu/wp-content/uploads/2014/02/graphx.pdf`.

YARN. `http://hadoop.apache.org/docs/current/hadoop-yarn/hadoop-yarn-site/index.html`.

Zaharia, Matei, Mosharaf Chowdhury, Tathagata Das, Ankur Dave, Justin Ma, Murphy McCauley, Michael J. Franklin, Scott Shenker, and Ion Stoica. *Resilient Distributed Datasets: A Fault-Tolerant Abstraction for In-Memory Cluster Computing*. `https://www.usenix.org/conference/nsdi12/technical-sessions/presentation/zaharia`.

Zaharia, Matei, Tathagata Das, Haoyuan Li, Timothy Hunter, Scott Shenker, Ion Stoica. *Discretized Streams: Fault-Tolerant Streaming Computation at Scale.* `https://people.csail.mit.edu/matei/papers/2013/sosp_spark_streaming.pdf`.

Zaharia, Matei. *An Architecture for Fast and General Data Processing on Large Clusters.* `http://www.eecs.berkeley.edu/Pubs/TechRpts/2014/EECS-2014-12.pdf`.

ZeroMQ. `http://zguide.zeromq.org/page:all.References`

Index

■ S

■ **T**

Get the eBook for only $5!

Why limit yourself?

Now you can take the weightless companion with you wherever you go and access your content on your PC, phone, tablet, or reader.

Since you've purchased this print book, we're happy to offer you the eBook in all 3 formats for just $5.

Convenient and fully searchable, the PDF version enables you to easily find and copy code—or perform examples by quickly toggling between instructions and applications. The MOBI format is ideal for your Kindle, while the ePUB can be utilized on a variety of mobile devices.

To learn more, go to www.apress.com/companion or contact support@apress.com.

Printed in the United States
By Bookmasters